Healthcare Kaizen in Action

Healthcare is a complex industry that requires efficiency, safety, and quality of care. Hospitals utilize structured problem-solving methodologies like the Plan–Do–Study–Act (PDSA) cycle and A3 thinking to achieve these goals. However, many healthcare professionals lack formal training in these methods.

Healthcare Kaizen in Action: 30 Case Studies for Rapid Improvement presents 30 real-life examples of Rapid Improvement Events (RIEs) in various healthcare settings. This book introduces the core principles of the PDSA cycle and A3 thinking and provides a detailed two-week timeline for implementing an RIE in a healthcare environment. Included are 30 collaboration briefs that detail specific improvement initiatives, such as reducing emergency department wait times, optimizing surgical scheduling, improving medication administration processes, and enhancing inventory management practices. Each brief follows a structured format, enabling readers to readily understand the case studies and apply the insights to their organizations. The case studies are organized by problem type, making it easy for readers to find solutions relevant to their challenges and identify actionable strategies to improve healthcare operations.

This book provides knowledge and tools for healthcare professionals to drive meaningful changes in their organizations, promoting continuous improvement and reducing inefficiencies. It is concise and actionable, aiming to improve patient outcomes and healthcare systems for nurses, physicians, hospital administrators, and process improvement specialists.

Healthcare Kaizen in Action
30 Case Studies for Rapid Improvement

Durward K. Sobek II and David Claudio

CRC Press
Taylor & Francis Group
Boca Raton London New York

CRC Press is an imprint of the
Taylor & Francis Group, an **informa** business

Designed cover image: Shutterstock

First edition published 2027
by CRC Press
2385 NW Executive Center Drive, Suite 320, Boca Raton FL 33431

and by CRC Press
4 Park Square, Milton Park, Abingdon, Oxon, OX14 4RN

CRC Press is an imprint of Taylor & Francis Group, LLC

ISBN: 978-1-041-13619-4 (hbk)
ISBN: 978-1-041-13534-0 (pbk)
ISBN: 978-1-003-67069-8 (ebk)

DOI: 10.1201/9781003670698

Typeset in Times
by codeMantra

To Maria, Angel, Amanda, and Anna. You remind me every day that continuous improvement isn't just for the workplace, it's a way of living, learning, and growing together.

And to our rural healthcare providers and support staff. Doing more with less is a way of life for you, and yet you always want to do better. You keep us practical, community-minded, and inspired.

Contents

Preface

Healthcare demands a balance of efficiency, safety, and quality, yet achieving all three remains a daily challenge. As hospitals and clinics work to reduce waste, waiting times, and errors, many are turning to structured problem-solving tools like the Plan–Do–Study–Act (PDSA) cycle and A3 thinking. Originally developed in manufacturing, these methods have proven equally powerful in healthcare, helping teams identify inefficiencies, implement rapid changes, and sustain improvements. However, few healthcare professionals receive formal training in these approaches, making them difficult to apply effectively.

This book bridges that gap by offering a practical, hands-on guide to Rapid Improvement Events (RIEs), which are structured, team-based efforts designed to effect measurable change. It begins with the fundamentals of PDSA and A3 thinking and outlines a concise two-week timeline for conducting an RIE. The heart of this book features 30 real-world case studies, each following a consistent format that makes it easy to apply lessons within your own organization. Topics include reducing emergency department wait times, improving surgical scheduling, optimizing medication administration, and enhancing inventory management.

Whether you are a nurse, physician, administrator, or part of a quality improvement team, this book is designed to be straightforward and actionable. You can pick it up and start using these tools right away. Our goal is simple: to help you build a culture of continuous improvement, one event at a time. By following the structured approach outlined here, you can make meaningful progress toward safer, more efficient, and more compassionate care for patients and the people who serve them.

About the Authors

Durward K. Sobek II is Vice Provost and Professor of Industrial and Management Systems Engineering at Montana State University. He is the co-founder of the Lean Product and Process Development Exchange Inc. He holds Ph.D. and M.S. degrees in industrial and operations engineering from the University of Michigan and an A.B. degree in engineering sciences from Dartmouth College. He has been researching lean product development and lean healthcare for over two decades, focusing on how organizations can increase their performance capacity through the application of lean principles.

David Claudio is Associate Professor and Program Director of the Industrial Engineering Program in the Department of Mechanical and Industrial Engineering at UMass Lowell, where he also serves as Associate Chair of the Industrial Engineering Graduate Programs. He holds a Ph.D. from The Pennsylvania State University, an M.S. from Rensselaer Polytechnic Institute, and a B.S. from the University of Puerto Rico, Mayagüez, all in industrial engineering. His research interests lie at the intersection of Healthcare Operations Management, Public Health, Human Factors and Ergonomics, Lean Principles applied to Service, and Predictive Modeling.

Acknowledgments

This book, and the ten years of work it represents, would not have been possible without the dedication, creativity, and persistence of the Industrial Engineering students (now professionals) who led and completed Rapid Improvement Events (RIEs) across hospitals and clinics throughout our program. Each case study, carefully selected and presented here, reflects their commitment to improving patient care, streamlining processes, and building stronger, more efficient healthcare systems. Their efforts not only made measurable improvements within each organization but also created lasting lessons that continue to inspire others to pursue excellence in healthcare operations.

To all the individuals who contributed their time, energy, and ingenuity to this work (ordered alphabetically by first name): Alex Bones, Amelia Farrier, Charlie Bartram, Dhaval Bajaria, Fletcher North, Jamie Schultz, Jeff Hickel, Jeremiah Coon, Joey Tschida, Jordan Burt, Kyle Kohn, Lauryn Yuzeitis, Matt Trzinski, Morgan Beckett, Naomi Pullmann, Ransom Taylor, Sam Allen, Stephen Manko, Tim Talbot, and Wyatt Justinak – thank you. Your problem-solving spirit and dedication to continuous improvement brought these RIEs to life, making this collection possible.

We would also like to express our sincere gratitude to the mentors who guided, challenged, and supported both the engineers and the program as a whole: Sean Harris, Kyle Kohn, and Jamie Schultz. Your leadership and commitment to lean principles, A3 thinking, and RIEs have not only shaped successful projects but also contributed to the development of the next generation of healthcare improvement professionals. Together, your contributions over the program's decade have left a lasting legacy of collaboration, learning, and impact across multiple healthcare organizations.

Many thanks to the Montana State Office of Rural Health, the Montana Hospital Association, and the Montana Health Research and Education Foundation for believing in us and underwriting much of the work. We especially thank the Flex Program Directors over the years, Carol Bischoff and Jamie Schultz, for their incredible partnership over the years and for providing so many introductions to facilities around Montana.

Finally, we would be remiss to not thank Bozeman Health and the critical access hospitals across the great state of Montana for warmly welcoming the interns, bravely embracing new ways of thinking and working, and being willing to experiment in the name of continuous improvement. Your tireless work to improve the lives of our rural communities is inspiring. Thank you.

1 Foundations of PDSA and A3 in Healthcare

Imagine you're dropped in the middle of the Pacific Ocean at night, under a new moon. You have no idea who put you there or why, no sense of where you are, and no clue how you're going to survive. They've tossed you a life vest, and now you have two choices: you can stay where you are, complaining, and slowly give up, or you can start swimming. But which way?

You don't know how to read the stars. You can't tell if the stars are moving or if the ocean is carrying you along. You can swim in any direction and hope to course-correct later, or you can wait and analyze the situation before taking action. Then you remember one thing: the sun always rises in the east. So you decide to wait until dawn and then swim east with the sunrise as your guide. That is the smart move because now you have direction, you have a goal, and you have a fixed point to navigate by. Instead of staying where you are, complaining, and slowly giving up, you're doing something about your situation; something well thought out and calculated.

Problems in healthcare systems often feel the same. We find ourselves dropped into messy situations, processes full of inefficiencies, waste, or confusion, and we do not always know the who, what, when, where, why, or how. But we do know this: we have options. We can complain until we burn out, or we can take action. Acting with a SMART goal in mind gives us direction, a target, and a way to measure progress. That is what this book is about: helping you find that direction and make meaningful progress.

Just as being lost in the ocean without a compass leaves you drifting aimlessly, healthcare organizations can face waves of problems without a clear way forward. A framework for improvement becomes that sunrise in the east, a reliable point of reference that allows movement with purpose. For this reason, many healthcare organizations have turned to structured problem-solving methodologies borrowed from industry. Among the most widely used are the A3 thinking and the Plan–Do–Check–Act (PDCA) cycle, often referred to in healthcare as Plan–Do–Study–Act (PDSA). These methods provide practical ways to identify problems, test solutions, and sustain improvements (Graban, 2016; Sobek & Smalley, 2011; Taylor et al., 2014).

Inside, you will find 30 real examples of Rapid Improvement Events (RIEs) or kaizen events. The details of each case are always unique to its setting, shaped by people, culture, and location, but the methods are consistent and repeatable. Your role is to do the detective work: to gather evidence, interview the people closest to the process, and assemble the facts into a clear story. Like a crime scene investigator piecing together a case, you bring order to chaos. That investigative discipline is what separates strong improvement leaders from the rest, and our goal is to equip you with a structured, methodical way to approach it, so you are not swimming aimlessly in the dark, but moving with purpose toward real, measurable change.

DOI: 10.1201/9781003670698-1

ORIGINS OF PDCA/PDSA IN HEALTHCARE

The PDCA cycle traces its origins to the work of Walter A. Shewhart, who initially conceptualized a repeating three-step improvement loop in the 1930s. The concept was then refined and formalized as a four-step cycle in the 1950s by W. Edwards Deming, who was once Shewhart's apprentice. Initially developed as a quality control method in industrial settings, the PDCA cycle promotes an iterative process: *plan* a change, implement it (*do* it!), evaluate (or *check*) the results, and *act* based on what is learned (Deming, 1986).

Deming later preferred the word "Study" over "Check" to emphasize the importance of methodical analysis and reflection during the evaluation phase, leading to the interchangeable PDSA terminology (Langley et al., 2009; Moen & Norman, 2010). Despite minor variations, the fundamental idea has remained the same; the PDCA/PDSA cycle is a structured method for continuous improvement that reflects the scientific method many of us first encountered in grade school, though easily forgotten in the complexity of daily work. From this point forward, we'll refer to it as the PDSA cycle, reflecting its widespread adoption as the standard terminology in healthcare improvement.

The adoption of PDSA in healthcare gained momentum in the late 20th century, particularly through the work of Don Berwick and the Institute for Healthcare Improvement (Berwick, 1989; Langley et al., 2009). The PDSA cycle now underpins the Institute for Healthcare Improvement's Model for Improvement and is widely used to guide quality initiatives such as reducing hospital-acquired infections, improving discharge processes, and increasing medication safety (Berwick, 1989; Christoff, 2018).

PDSA encourages healthcare teams to test changes on a small scale, study results, and iteratively refine interventions. This approach to operational improvement has been particularly effective in addressing issues ranging from medication errors and infection control to staff workflow and patient wait times (Christoff, 2018; Reed & Card, 2016; Sobek et al., 2012; Hickel et al., 2012).

While PDSA is widely adopted, its application often lacks rigor. Taylor et al. (2014) observed that many healthcare organizations use PDSA without clearly defined predictions or multiple iterative cycles. To strengthen improvement efforts, Glover et al. (2014) recommended conducting RIEs more frequently to create better conditions for success. An example comes from Nino et al. (2020), who carried out three consecutive PDSA cycles within a single RIE, focusing first at the departmental level, then at a specific area, and finally at the workstation level (see Figure 1.1).

This approach illustrates the concept of the improvement ramp (Bradley et al., 2024; Iwao, 2018; Glover et al., 2014; Sobek & Smalley, 2011; Nelson et al., 2011), where successive, nested PDSA cycles lead to progressively refined outcomes. Within 2 weeks, the cycles yielded a streamlined workflow and a redesigned sterile processing department layout, including one of its areas and an individual workstation. Reed and Card (2016) argue that PDSA should be treated as a structured learning tool, not just a checklist. In other words, similar to a continuous improvement mindset, PDSA is a way of life rather than something we do.

To be effective, teams must make explicit predictions, measure outcomes, and learn from failures (Christoff, 2018). When applied with intention and discipline,

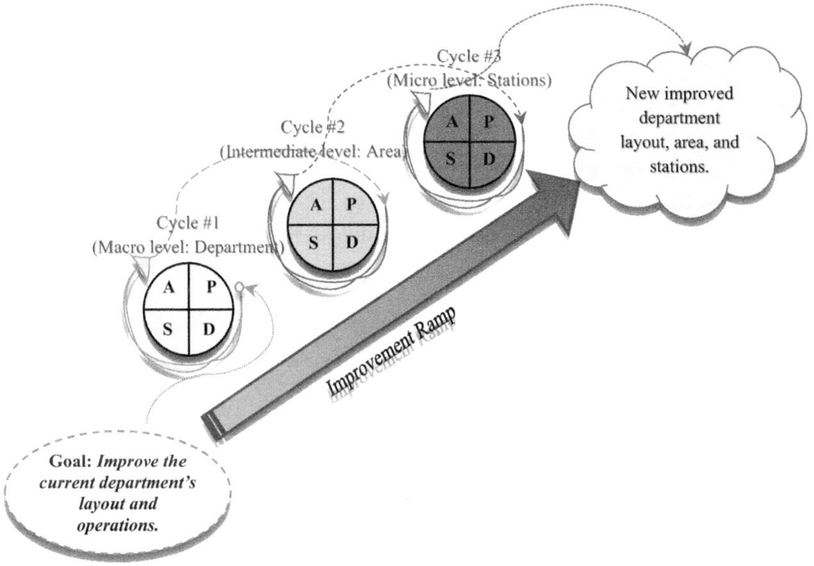

FIGURE 1.1 Successive PDSA cycles applied within a single RIE, illustrating the improvement ramp as changes progressed from the departmental level to a specific area, and finally to an individual workstation.

PDSA enables rapid, small-scale testing that reduces the risk of full-scale implementation failures.

ORIGINS AND STRUCTURE OF A3 THINKING

The A3 report was developed within the Toyota Production System as a communication tool to summarize problem-solving on a single A3-sized (roughly 11″ × 17″) sheet of paper. It promotes clarity, structured thinking, and alignment among teams (Liker, 2004; Sobek & Smalley, 2011). More than a document for communication purposes, the A3 report is a learning tool that captures the PDSA cycle in visual and logical form as well as a collaboration tool that facilitates consensus decision-making and mentoring. Sobek and Smalley (2008) suggest that the thinking behind A3 reports is more important than the report itself and characterize A3 thinking with seven elements:

Logical Thinking Process: A3 thinking follows a cause-and-effect structure that mirrors the scientific method. Each step builds logically onto the previous one, preventing teams from jumping to solutions before developing a deep, shared understanding of the problem or opportunity. This creates a straightforward storyline that others can easily follow from start to finish.

Objectivity: A3 thinking emphasizes grounding decisions in facts and data gathered from the actual workplace (the so-called "Gemba"). This keeps problem-solving tied to reality rather than assumptions. Decisions are built on what you see and measure, not on opinions, guesses, or assumptions. You go to the workplace, walk through the process, talk with people, and use real data.

Results and Process: A3 thinking highlights both the results achieved and the process that leads to them. The "what" and the "how" are equally important because strong processes lead to repeatable and lasting improvements because the complexities are well understood and there is consensus on the best (at least for now) way forward. It's not just about solving a problem; it's about building a way of working that can be learned from and applied again in the future.

Synthesis, Distillation, and Visualization: By fitting the whole story on a single A3 sheet, you're forced to boil things down to the essentials. That means using visuals and short explanations instead of long reports. This discipline enhances clarity and communication. It is the application of the KISS principle: keep it short and simple.

Alignment: An A3 report should not be written alone. Rather, it is used to gather and synthesize information from multiple sources, to share ideas, get input and new ideas, and solicit support. It's meant to get everyone on the same page (figuratively and literally!). Used in conjunction with collaborative problem-solving practices, it aligns team members, departments, and leaders around a shared understanding of the problem, goals, and countermeasures.

Coherency Within and Consistency Across: Each part of the A3 report should line up with the others. The problem connects to the root causes, the countermeasures address those causes, and the results show whether the plan worked. Consistency is attained when all parts of the organization use a common tool and approach for problem-solving, resulting in common vocabulary and processes that facilitate greater efficiencies when working across organizational boundaries.

Systems Viewpoint: Solving one problem should not create a bigger one somewhere else. A3 thinking encourages you to see how your changes fit into the larger system. Rather than optimizing a single part in isolation, the process asks how a solution impacts the entire workflow or organization. Figure 1.2 presents seven building blocks of A3 thinking in a simplified manner. Each piece works together to give problem-solvers clarity, teamwork, and a big-picture view.

A3 thinking entered healthcare as part of Lean management adoption, particularly through systems like Virginia Mason Medical Center and ThedaCare (Graban, 2016). A3 reports have since been used in a wide range of clinical and operational contexts, from reducing hospital-acquired infections to streamlining discharge planning. For example, Lee and Kuo (2009) demonstrated how A3 could be used to improve hospital workflows by identifying bottlenecks and aligning team interventions. Additionally, A3s have been used as strategic tools in performance management. Healthcare organizations use "cascading A3s" to align front-line projects with enterprise-level goals (Graban, 2016).

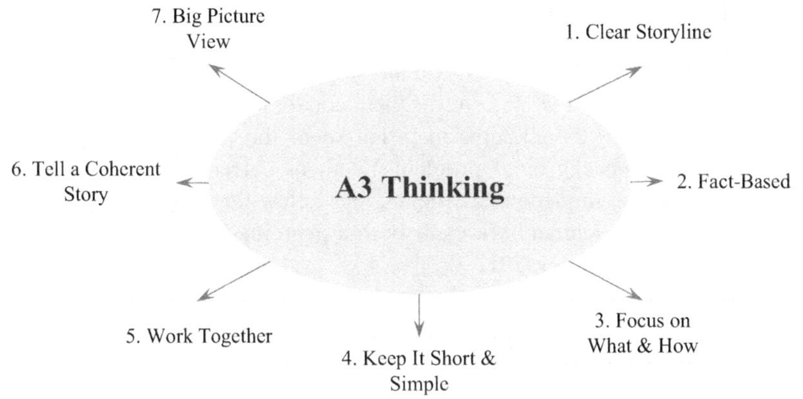

FIGURE 1.2 The seven building blocks of A3 thinking shown as parts of one system.

PDSA AND A3 REPORTS AS COMPLEMENTARY TOOLS FOR CONTINUOUS IMPROVEMENT

A3 reports and PDSA aim for iterative, data-driven improvement, and are rooted in the Deming cycle of PDCA. The difference is largely in scope and structure. A PDSA form (or worksheet) is typically used to plan a single experiment or test of change. It guides you to state a hypothesis, run a small test, observe the results, and decide next steps. In healthcare quality improvement projects, teams often cycle through multiple PDSA forms as they tweak and refine an intervention (for example, testing different ways to reduce wait times, 1 week at a time).

An A3 report, on the other hand, provides a bigger picture framework. It encompasses not just one test, but the narrative of the entire problem-solving effort, from problem definition through implementation and follow-up. In fact, the A3 process is one large PDSA cycle. Approximately 70% of the A3 process (Background, Current State, Analysis, Target, Root Cause Analysis, Proposed Countermeasures) is essentially the planning phase (Plan). The last steps (Implementation, Follow-up) cover Do, Study, and Act. Many organizations find that using the A3 thinking process in tandem with running quick PDSA cycles gives the best of both worlds: the A3 brings the discipline of thinking through the whole problem and keeping track of progress, while PDSA cycles allow nimble experimentation within that structure. In fact, you can incorporate several PDSA cycles within an A3 process as part of your countermeasures and implementation plan, essentially nesting multiple smaller PDSA cycles within the larger PDSA cycle of the A3.

One practical difference is format and communication. PDSA forms can be a bit siloed; you might fill them out and file them away for the quality improvement team's use. An A3 report, being a story-telling document, is much easier to share and discuss broadly. It's common to review A3s in staff meetings or post them on a wall, whereas a stack of PDSA forms wouldn't communicate as clearly. In essence, an A3 report can serve as a high-level consolidation of multiple PDSA cycles, providing a summary that stakeholders can quickly grasp. The Public Health Foundation and other organizations have published A3 templates explicitly aligned to PDSA steps (Lean Enterprise Institute, Inc., n.d.b), underscoring that the two go hand-in-hand.

In summary, think of PDSA forms as the tactical tools for rapid experimentation, and the A3 process as the strategic map that keeps the whole effort coherent. The A3 process includes background, problem statement, current state, root cause analysis, countermeasures, implementation plan, and follow-up evaluation. It mirrors the stages of the PDSA cycle and serves as both a planning document and a record of outcomes (Sobek & Smalley, 2011).

The combination of PDSA and A3 has laid the groundwork for what many call a "culture of continuous improvement" or kaizen in healthcare (Lean Enterprise Institute, Inc., n.d.a). A3 thinking provides a deeper level of granularity to the PDSA cycle, and its final product becomes the artifact that documents each PDSA cycle, helping teams learn from past cycles and maintain transparency in progress (Sobek & Smalley, 2011; Reed & Card, 2016).

Together, PDSA and A3 thinking form a powerful duo: PDSA provides the iterative improvement method, while A3 provides a format for organizing and communicating the work. Used in tandem, they support disciplined, evidence-based improvement efforts that build shared understanding across teams (Reed & Card, 2016; Sobek & Smalley, 2011). These tools are not just about fixing problems. They are about developing people and systems that can continuously learn and improve. By applying this combined approach, organizations foster a culture of inquiry, transparency, and accountability at every level.

This book brings these concepts to life through 30 case studies. Each one will show how PDSA and A3 thinking were used in real-world scenarios to address pressing healthcare challenges. The case studies include real problems from scheduling and inventory issues to workflow inefficiencies and charge capture. These stories will demonstrate that improvement is not only possible but practical, even in the complexity of healthcare systems in the 21st century.

A3 THINKING VS. SIX SIGMA'S DMAIC

Another widely used improvement approach is Six Sigma (George et al., 2005). In this approach, teams follow a high-level improvement cycle summarized by the DMAIC acronym, which stands for Define, Measure, Analyze, Improve, and Control. If you compare DMAIC to the A3 structure, you'll see a strong alignment:

Define = Background/Problem Definition
Measure = Current State data
Analyze = Root Cause Analysis
Improve = Countermeasures/Implementation
Control = Follow-up sustainability

In other words, A3 and DMAIC cover very similar steps; they're just rooted in different traditions (Lean for A3, and statistical quality control for Six Sigma) (Lee et al., 2025; Yang et al., 2025; Nino et al., 2021; Taylor et al., 2014; Reis et al., 2022). One key difference is emphasis. Six Sigma projects (especially classic DMAIC ones) tend to invest heavily in the Measure and Analyze phases, using detailed data collection and statistical analysis to pinpoint causes of variation. Lean's A3 tends to emphasize understanding workflows and eliminating waste, often with simpler visuals and time or capacity analyses, and encourages using just enough analysis to get to the root cause. However, these approaches are increasingly blended into practice. It's not unusual to see a "Lean Six Sigma" (Abualsauod, 2025; George et al., 2005) project where an A3 is used to storyboard a DMAIC initiative. In fact, the structured format of A3 can bring clarity to DMAIC projects by ensuring that the problem is well-defined and the improvements are tied back to the identified causes.

Just as PDSA and A3 are complementary, so are DMAIC and A3. If you're working in a Six Sigma environment, you can use an A3 report to document your project as you move through DMAIC phases. Conversely, suppose you're coming from a Lean perspective but need deeper statistical insight. In that case, you can use a DMAIC cycle or parts thereof in your A3 (for example, using a Pareto chart or control chart in the analysis section). Both approaches value a structured, logical progression from problem to solution. A3 thinking provides a communication and collaboration vehicle, while DMAIC provides a toolset for analysis and improvement. The choice isn't necessarily one or the other, but when and how to use both effectively.

THE ROLE OF SMART GOALS IN PDSA AND A3 THINKING

A key enabler of effective PDSA and A3 implementation is the use of SMART goals. These are goals that are *S*pecific, *M*easurable, *A*chievable, *R*elevant, and *T*ime-bound (Doran, 1981). Originally developed as a management tool in organizational planning, SMART goals have become a cornerstone of successful quality improvement in healthcare, particularly in the Plan stage of PDSA. In fact, several healthcare improvement frameworks, including those promoted by NHS England and the Institute for Healthcare Improvement, explicitly recommend incorporating SMART goal setting at the outset of improvement cycles (NHS England, n.d.; Langley et al., 2009). Setting a vague or overly ambitious aim (e.g., "make discharge faster") often leads to frustration or unfocused effort. In contrast, a SMART goal like "reduce average patient discharge time by 30%, from 4 hours to 2.8 hours, within the next 90 days" provides a clear benchmark and timeline that anchors the PDSA cycle in measurable, actionable terms (ACP Decisions, 2019).

According to the Public Health Foundation's A3 template, improvement objectives should clearly outline what the team hopes to achieve, by when, and how success will be measured, directly aligning with the SMART framework (PHF, 2025). Whether a team is trying to reduce medication errors, increase staff compliance with hand hygiene, or streamline a scheduling process, a SMART goal creates focus and accountability throughout the improvement journey.

Beyond planning, SMART goals also shape how success is evaluated in the Check/Study and Act phases of PDSA. Without a clear baseline and target, teams may not know whether their intervention worked. When an improvement project includes SMART goals, the final evaluation becomes more objective and meaningful: the team can show not just that a change was made, but whether it closed the gap between the current state and the target condition.

Just as SMART goals bring clarity and focus to improvement work, we've applied the same principle to this book. Our intention is not only to share knowledge, but to do so in a way that leads to practical, measurable outcomes for you, the reader. Let's break down our goals for this book using the SMART framework.

By the time you finish this book, we aim to accomplish the following:

1. **Document and present 30 complete real RIE case studies**, categorized by improvement domain and formatted for easy adaptation across diverse healthcare settings.
   Specific: 30 RIE case studies
   Measurable: Counted and categorized
   Achievable: Drawn from actual projects
   Relevant: Supports learning and replication
   Time-Bound: Completed before publication
2. **Keep theoretical content (first three chapters) to no more than 25% of the book**, ensuring that at least 75% is devoted to real-world case studies drawn from healthcare RIEs.
   Specific: Limit theory
   Measurable: Max 25% of content in the first three chapters
   Achievable: Built into book structure
   Relevant: Focuses on practical application
   Time-Bound: Measured by the final document (before editing by the publisher)
3. **Ensure that within 1 year of reading this book, you, the reader, have applied at least one PDSA cycle or RIE** in your own workplace using the guidance and examples from this book.
   Specific: Apply one improvement cycle
   Measurable: Reader self-report
   Achievable: Supported with tools and templates
   Relevant: Aligns with readers' improvement roles
   Time-Bound: 1 year from reading
4. **Enable readers to participate in a team-based improvement initiative using the A3 framework within 6 months of completing the book**.
   Specific: Participate in an A3-based project
   Measurable: One initiative per reader
   Achievable: Aligned with case studies and guidance provided
   Relevant: Builds core Lean leadership capability
   Time-bound: 6 months after reading

These goals reflect not just our content strategy, but our broader intention: to equip you with tools and insights you can use. If we've met these goals, you'll walk away not only understanding PDSA and A3 thinking, but having used them and perhaps shared them with others.

REFERENCES

Abualsauod, E. H. (2025). Integrating Six Sigma and Lean management for enhanced quality management in industrial applications. *Journal of Engineering Research.* Advance online publication. https://doi.org/10.1016/j.jer.2025.09.007

ACP Decisions. (2019, September 30). *5 tools to maximize the effectiveness of your PDSA cycles.* ACP Decisions. Retrieved July 27, 2025, from https://www.acpdecisions.org/5-tools-to-maximize-the-effectiveness-of-your-pdsa-cycles/

Berwick, D. M. (1989). Continuous improvement as an ideal in health care. *New England Journal of Medicine*, 320(1), 53–56. https://doi.org/10.1056/NEJM198901053200110

Bradley, C., Sumethasorn, M., Kim, J. S., Wang, S., Martinez, L., Chang, M., Lemus, L., Katebian, B., Bruce, D., Lee, A., Baden, R., Yee, H., & Buxbaum, J. (2024). Plan-Do-Study-Act (PDSA) interventions to improve real-world endoscopy unit productivity. *Endoscopy International Open*, 12(5), E642–E648. https://doi.org/10.1055/a-2290-0263

Christoff, P. (2018). Running PDSA cycles. *Current Problems in Pediatric and Adolescent Health Care*, 48(8), 198–201. https://doi.org/10.1016/j.cppeds.2018.08.006

Deming, W. E. (1986). *Out of the crisis.* MIT Press.

Doran, G. T. (1981). There's a S.M.A.R.T. way to write management's goals and objectives. *Management Review*, 70(11), 35–36.

George, M. L., Maxey, J., Rowlands, D. T., & Upton, M. (2005). *The Lean Six Sigma pocket toolbook.* McGraw-Hill, 1–85.

Glover, W. J., Farris, J. A., & Van Aken, E. M. (2014). Kaizen events: Assessing the existing literature and convergence of practices. *Engineering Management Journal*, 26(1), 39–61. https://doi.org/10.1080/10429247.2014.11432003

Graban, M. (2016). *Lean hospitals: Improving quality, patient safety, and employee engagement* (3rd ed.). CRC Press. https://doi.org/10.4324/9781315380827

Hickel, J., Talbot, T., Claudio, D., & Sobek, D. K. (2012). A case study on reducing late first-round surgery OR arrivals. *Proceedings of the 2012 Industrial and Systems Engineering Research Conference*, Orlando, FL.

Iwao, S. (2018). The diversity and reality of Kaizen in Toyota. In Iwao, S. (Ed.), *Industrial competitiveness and design evolution* (pp. 271–298). Springer.

Langley, G. J., Nolan, K. M., Nolan, T. W., Provost, L. P., & Norman, C. L. (2009). *The improvement guide: A practical approach to enhancing organizational performance* (2nd ed.). Jossey-Bass.

Lean Enterprise Institute. (n.d.a). *Plan-Do-Check-Act (PDCA).* Lean.org. Retrieved July 20, 2025, from https://www.lean.org/lexicon-terms/pdca/

Lean Enterprise Institute. (n.d.b). *Problem-solving templates: PDSA A3 template (from On the Mend).* Retrieved August 19, 2025, from https://www.lean.org/events-training/forms-templates/

Lee, T. S., & Kuo, M. H. (2009). Toyota A3 report: A tool for process improvement in healthcare. *Studies in Health Technology and Informatics*, 143, 235–240. https://pubmed.ncbi.nlm.nih.gov/19380942

Lee, T., McMartin, L. J., & Woodworth, J. (2025). Basics of Lean and Six Sigma: Fundamental tools for health care providers. *Obstetrics and Gynecology Clinics of North America.* Advance online publication. https://doi.org/10.1016/j.ogc.2025.09.002

Liker, J. K. (2004). *The Toyota way: 14 management principles from the world's greatest manufacturer.* McGraw-Hill.

Moen, R., & Norman, C. (2010). Circling back: Clearing up myths about the Deming cycle and seeing how it keeps evolving. *Quality Progress*, 43(11), 22–28.

Nelson, E. C., Batalden, P. B., & Godfrey, M. M. (2011). *Value by design: Developing clinical microsystems to achieve organizational excellence.* John Wiley & Sons.

NHS England. (n.d.). *PDSA cycle template and learning guide*. NHS England. Retrieved August 3, 2025, from https://www.england.nhs.uk/improvement-hub/wp-content/uploads/sites/44/2015/08/learning-handbook-pdsa.pdf

Nino, L., Claudio, D., Valladares, L., & Harris, S. (2020). An enhanced Kaizen event in a sterile processing department of a rural hospital: A case study. *International Journal of Environmental Research and Public Health*, 17(23), 8748. https://doi.org/10.3390/ijerph17238748

Nino, V., Martinez-Torres, K., Gomez-Guzman, K., & Claudio D. (2021). Improving the registration process in a healthcare facility with lean principles. *Journal of Industrial Engineering and Management*, 14(3), 538–551. https://doi.org/10.3926/jiem.3432

Public Health Foundation (PHF). (2025). *A3 problem solving report template*. Public Health Foundation. Retrieved August 3, 2025, from https://phf.org/wp-content/uploads/2025/06/A3_Problem_Solving_Report.pdf

Reed, J. E., & Card, A. J. (2016). The problem with Plan-Do-Study-Act cycles. *BMJ Quality & Safety*, 25(3), 147–152. https://doi.org/10.1136/bmjqs-2015-005076

Reis, M. E. D. M. dos, Abreu, M. F. de, Braga Neto, O. de O., Viera, L. E. V., Torres, L. F., & Calado, R. D. (2022). DMAIC in improving patient care processes: Replication and lessons learned in the context of healthcare. *IFAC-PapersOnLine*, 55(10), 549–554. https://doi.org/10.1016/j.ifacol.2022.09.451

Taylor, M. J., McNicholas, C., Nicolay, C., Darzi, A., Bell, D., & Reed, J. E. (2014). Systematic review of the application of the Plan-Do-Study-Act method in healthcare. *BMJ Quality & Safety*, 23(4), 290–298. https://doi.org/10.1136/bmjqs-2013-001862

Sobek, D. K., Claudio, D., & Bischoff, C. (2012). Rapid improvement events across rural hospitals in the state of Montana. *Proceedings of the 2012 Industrial and Systems Engineering Research Conference*, Orlando, FL.

Sobek, D. K., & Smalley, A. (2011). *Understanding A3 thinking: A critical component of Toyota's PDCA management system*. CRC Press. https://doi.org/10.4324/9781439814055

Yang, L.-L., Zhang, H., Zhang, S.-H., Gao, P., Yan, S.-T., & Li, D.-N. (2025). Implementing Lean Six Sigma to optimize emergency extracorporeal cardiopulmonary resuscitation process management. *The American Journal of Emergency Medicine*, 98, 33–40. https://doi.org/10.1016/j.ajem.2025.08.010

2 The A3 Thinking Process

Imagine it's 7:30 a.m. in a busy hospital, and the first surgery of the day is already running behind schedule. Nurses, anesthesiologists, and the surgeon are scrambling because the patient wasn't in the operating room (OR) at the 7:15 scheduled time. You can feel the tension; everyone knows that when the first case starts late, the delays cascade throughout the day. What can you do to prevent this situation from happening tomorrow and the day after? This is where A3 thinking comes into play.

A3 thinking is a problem-solving approach named after the A3 paper size (roughly 11″ × 17″), which helps you and your team systematically identify the cause of a problem, devise countermeasures, and follow through with an improvement plan, all summarized on one piece of paper. Traditionally, that single-sheet format is what gives the A3 report its power: it forces brevity, clarity, and focus.

In our work running Rapid Improvement Events (RIEs) in healthcare settings, however, we adapted the classic format into what we call an RIE brief. The template sections are slightly different, tailored to the RIE context, and we do not constrain the report to a single A3 page or to two 8.5″ × 11″ pages. In fact, 90% of the RIE briefs in this book are between four and eight pages long. The reasons for this are several. First, the most important part of the A3 approach is the thinking behind it, not the length of the report. As long as we follow the A3 process and use the report in the spirit of A3 thinking, we feel that the report's length is secondary. Second, in practice, developing an A3 report is an iterative process in which multiple versions of the report are drafted as new learning and discoveries emerge. Often, a later iteration must summarize earlier versions to make room for new learning and stay within the imposed size constraints. Because RIEs are on a fixed, tight timeline, we found that the effort to develop several iterations of an A3 report took too much valuable time away from the problem-solving and collaborative efforts required for success. So rather than summarizing earlier work to create space, we coached our facilitators to add new findings to it. In addition, preserving the details in the earlier steps of the process allowed the document to be used as a training tool for new team members and staff (and for readers of this book) and to capture the reasoning behind the decisions, which is essential for effective change management and future continuous improvement efforts. The expanded format honors the thinking behind the A3 approach while being a better fit to the specific needs of RIEs.

This adaptation is also what inspired us to put this book together. By adapting the A3 approach into a shareable RIE brief format, we realized we could do more than solve problems inside our organizations. We could capture the stories, lessons, and methods in sufficient detail to share them with others. This book is our way of opening that door, so that, regardless of your role, you can learn from these experiences and apply them in your own setting.

DOI: 10.1201/9781003670698-2

In this chapter, we'll walk you through the A3 thinking process. We'll also put theory into practice with a real-world example of tackling first-case surgery delays, step by step, using the A3 process and our RIE brief format.

A3 THINKING STRUCTURE

At a basic level, the A3 report serves as a storyboard for your improvement project. It captures the key elements of your Plan–Do–Study–Act (PDSA) cycle in a concise, visual format. The report is divided into sections that logically flow from defining the problem through implementing and checking a solution, coinciding with the A3 thinking process as described by Sobek and Smalley (2011) and presented in Figure 2.1. You can see how information flows from defining the problem through analysis to results and follow-up. It also includes collaboration steps to discuss learning, findings, and ideas with those potentially affected by the problem and/or potential solutions, and appropriate approvals.

The RIE briefs presented in this book follow the same order of the A3 thinking process, as illustrated by the RIE brief template shown in Figure 2.2. The rest of this section explains the template by walking through the A3 thinking process, starting with perceiving a problem.

Problem Perceived: Before beginning the A3 process, the hospital department or clinic sponsor must identify and propose specific improvement opportunities or problems they have observed. They should then define clear objectives and prepare relevant data prior to the event, while project sponsors coordinate with stakeholders

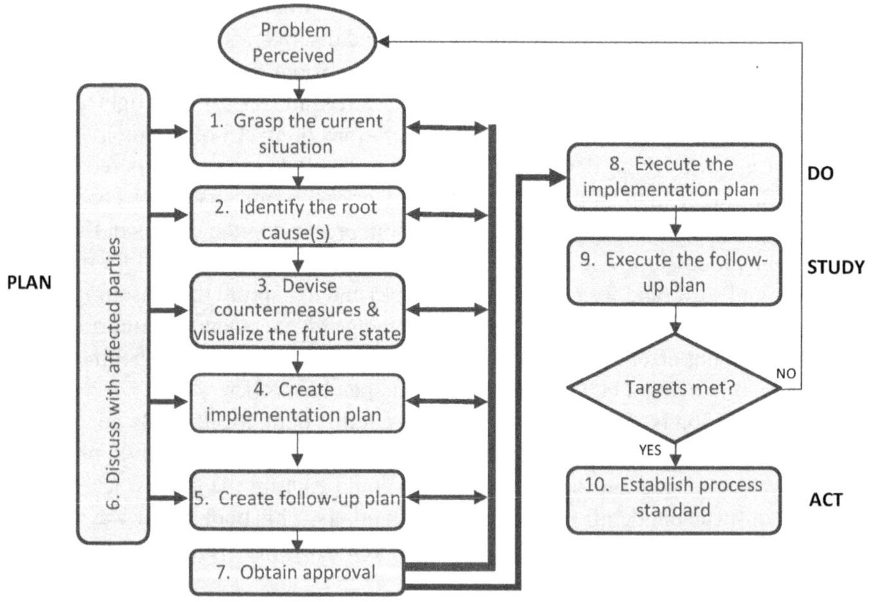

FIGURE 2.1 The A3 problem-solving processes. (Extracted from Sobek & Smalley, 2011, page 20).

Theme:	Date:
Sponsor:	Author(s):

Background:

-
-

Current Condition:

SMART Goal(s):	➢

Root Cause Analysis:

Proposed Countermeasures
Target Condition:

Implementation Plan

Action	Responsible	Deadline	Expected Outcome	Actual Outcome

Effects Confirmation

Follow-up Actions

Action	Responsible Party	To Complete Action	Due

FIGURE 2.2 The RIE brief template grounded in A3 process thinking.

to ensure participation and readiness. This structured collaboration creates the foundation for generating measurable improvements in care delivery.

1. **Grasp the Current Situation**: The first elements of the RIE brief template correspond to the first step in the A3 process.
   - **Theme**: More than a simple title, the report theme describes what the RIE is trying to achieve, and does so without presuming a solution. (E.g., Reducing Patient Discharge Time in the ER versus Upgrade the Pager System.)
   - **Sponsor**: It is important to indicate the individual responsible for the success of the RIE. Sometimes it is tempting to list a department or team as the sponsor, but generally it is preferable to identify a person who takes ownership of the project's success.
   - **Date and Author**: This is important to understand how changes have been applied to a process or system over time and who has led the RIE efforts (facilitator).

- **Background**: This section of the RIE brief provides context and explains the rationale for selecting the problem at hand. You explain why this issue is important. It may also frame how the problem aligns with organizational values or strategic goals.
- **Current Condition**: Here, you factually describe what is happening now. This often includes baseline data, process maps, or observations from the Gemba (the workplace) and the Gemba walk. The goal is to quantify the problem and illustrate how the process currently works. Key performance metrics, timelines, or any visual representations (such as a flow diagram with problem areas highlighted) are useful. Essentially, the current state paints a clear and accurate picture of reality, including the extent of the problem (e.g., error rates, delay times, breakdowns in communication) and any obvious waste or inefficiencies.
- **SMART Goal(s)**: Here is where the facilitator states the RIE objectives in a Specific (what we're improving), Measurable (a quantitative metric), Achievable (within the RIE timeframe), Relevant (important to the organization), and Time-bound (for RIEs, we will always work under the statement: By the end of this RIE, we will …).

2. **Identify the Root Causes**: In an RIE brief, root causes are identified in the Root Cause Analysis section. At this step, you dig deeply into why the problem is occurring. The team identifies potential causes and uses analytical tools to find root causes, often employing the 5 Whys technique (Sobek & Smalley, 2011; AHRQ, 2015), fishbone diagrams (Kumah et al., 2024; AHRQ, 2015), and other quality tools. A root cause is one that, if properly addressed, will prevent the problem from recurring. The A3 process encourages using the simplest tool that suffices to reveal cause-and-effect, but it doesn't shy away from more sophisticated analysis if needed. The output here is a clear identification of the underlying cause(s) of the gap between current and desired performance.

   Warning: Please avoid blaming individuals for mistakes (e.g., employees not completing their work) as the ultimate cause. Instead, look for systematic failures (i.e., why are the employees not doing their work? → Because no one has explained to them their roles and responsibilities. Why?…)
3. **Devise Countermeasures and Visualize the Future State**: This step is represented in the Proposed Countermeasures section. Here you list or illustrate possible ways to address the root causes, i.e., countermeasures. We use the word "countermeasures" rather than "solutions" to convey the idea that we are going to test these ideas and measure their effectiveness before we make them standard practice.
   - **Proposed Countermeasures**: Each countermeasure should directly relate to a cause you have identified. You might propose multiple countermeasures, but they should form a coherent strategy to reach the target condition. It's good practice to briefly justify each countermeasure, explaining why this action will help address the problem. Brainstorm potential ideas with the people involved, especially those actually doing the work, to identify and address the root causes of the issue.

- **Target Condition**: This section defines what "good" looks like once the countermeasures are in place. In other words, your goal or expected future state after improvements. It may also include a mini vision of the new process, sometimes accompanied by a diagram of how things should flow after the changes.

4. **Create an Implementation Plan**: The implementation plan for an RIE essentially serves as an action plan, outlining what needs to be done, when, and by whom for the countermeasures to be developed and tested. The plan should be realistic and account for necessary resources or approvals. We adopted a table format that lists actions, responsible owners, deadlines (or due dates), expected outcomes, and actual outcomes for each action. These last two columns allow us to document whether the action had the intended effect.
5. **Create a Follow-Up/Assessment Plan**: The follow-up plan in the A3 process specifies how you will measure and verify the results once countermeasures are implemented. It might include a list of metrics to check, dates for post-implementation review, and assignment of who will do the monitoring. After implementation, the team records the actual results here and notes any further actions (e.g., adjustments, standardization, or spread of the solution). Essentially, this part answers: did we achieve the target? Did our countermeasures work, and what should we do next (standardize the improvement, iterate further, etc.)?

   In an RIE brief, you will often find an action item in the implementation plan that includes these actions, perhaps as a pilot study. You will also see under Effects Confirmation how changes were measured and the impact. Sometimes, a picture is evidence enough that a checklist, sign, or form has been created, or if an area has changed its layout. At other times you will find graphs or plots, such as those collected at the current condition state, but for the new process.
6. **Discuss with Affected Parties**: A defining feature of the A3 process is the engagement of everyone who may be affected by the improvement work, especially front-line stakeholders. The same goes for RIEs. Continuous Quality Improvement (CQI) and other process improvement professionals have learned that improvements are most effective and sustainable when they are designed and implemented by the people who do the work, with support from leadership, rather than imposed top-down. Although labeled as "Step 6," discussion with affected parties should be ongoing at every stage of the process. For example, once the current condition is sufficiently studied and documented, the RIE brief is reviewed with the people involved in the work to verify that it accurately represents what is currently happening. The facilitator should incorporate any input received during this discussion into the current condition to develop a shared understanding of the current situation. This naturally leads to a discussion on defining the goals for the RIE. In a similar fashion, the proposed countermeasures and envisioned future state is also discussed with stakeholders to solicit input, incorporate ideas and suggestions, and collectively agree on the best course

of action. Like A3 reports, discussing with affected parties is not explicitly in the RIE brief, although the presence of an implementation plan implies that the countermeasures, responsible parties, and deadlines have been discussed with the appropriate stakeholders.

7. **Obtain Approval**: Like the previous step, obtain approval is not explicitly mentioned in the RIE brief. Similarly, this is another step that occurs on an ongoing basis through the previous steps. You should obtain approval from the sponsor and other appropriate leaders as early as when scoping the problem and creating the SMART goals. However, even if you have done everything necessary and possible to obtain buy-in and approval from the affected parties, from our experience, you might find that some people change their minds, get scared of change, or backpedal when it is time to implement the plan (the Do stage of the PDSA). Since the 30 reports presented in this book were developed from real RIEs, you will find a few instances in which at least one person (or management) did not support a countermeasure. In some instances, we have reported this under the "Actual Outcome" column of the implementation plan or notified management via a note in the Effects Confirmation section.
8. **Execute the Implementation Plan**: Up to this point, you have only done the "Plan" stage of the PDSA cycle. Now it is time to execute the "Do" stage of the PDSA. As the implementation plan outlined in the RIE brief is executed, the report author fills in the Actual Outcomes column to indicate whether that action was taken, any modifications to the original plan, and the result of the action.
9. **Execute the Follow-Up Plan**: In the A3 process, this is the "Study" stage of the PDSA.
   - **Effects Confirmation**: You report the "study" part by presenting evidence on how the new system state (with changes) compares to the old one (formerly known as the current condition or current state). At this stage, you also state which SMART goals have been met by providing data on the metric(s) associated with each goal.
   - If one or more SMART goals were not met, it is still recommended to state any improvement that was made. Every incremental improvement counts (as opposed to doing nothing); you should still celebrate small wins. In a few of our RIE briefs, you will find that the SMART goal was not met. For example, in Chapter 6, the RIE brief on "Improving Workflow in Medical Center Emergency Room," one of the SMART goals was to "reduce search time standard deviation by 50%." However, the standard deviation was reduced by 43.48%, 6.52% shy of the goal. We still celebrated the reduction in standard deviation, as the goal may have been too ambitious or may have been met with a larger sample size (with more time). At the same time, we took the opportunity to reflect on why we thought the goal was achievable and why we did not achieve it.
   - If goals are not improved at all, then you should revisit the beginning and revise the project scope, assumptions, root cause analysis, or countermeasures.

10. **Establish Process Standard**: In A3 thinking, establishing the new way of working as a standard procedure is the "Act" stage of the PDSA. The entire RIE brief tells the story of how the process used to be and the new processes that were collaboratively developed by the affected parties. In the case of RIEs, the time is often too short to fully establish the new way of working. That happens in the days and weeks following the event. Thus, the RIE brief outlines the steps that need to be taken to establish that process standard and verify that it is having the intended effect. The RIE brief lists these steps in a Follow-up Actions section, not to be confused with the follow-up/assessment step in the A3 process.
    - **Follow-Up Actions**: Follow-up and sustainment activities are crucial for transforming the event's gains into lasting improvements. Without diligent post-event work, there is a risk that people revert to old habits or that implemented changes are not fine-tuned and thus fall short of expected results. It's also worth noting that not every RIE will achieve all its goals or be able to implement the countermeasures within the RIE timeframe. That's okay if it leads to learning and further cycles of improvement.

To further illustrate how RIE briefs incorporate A3 thinking, we have revised the A3 process presented in Figure 2.1 to align with the RIE brief as presented in Figure 2.3. Additionally, we overlaid the A3 process steps on the RIE brief template in Figure 2.4.

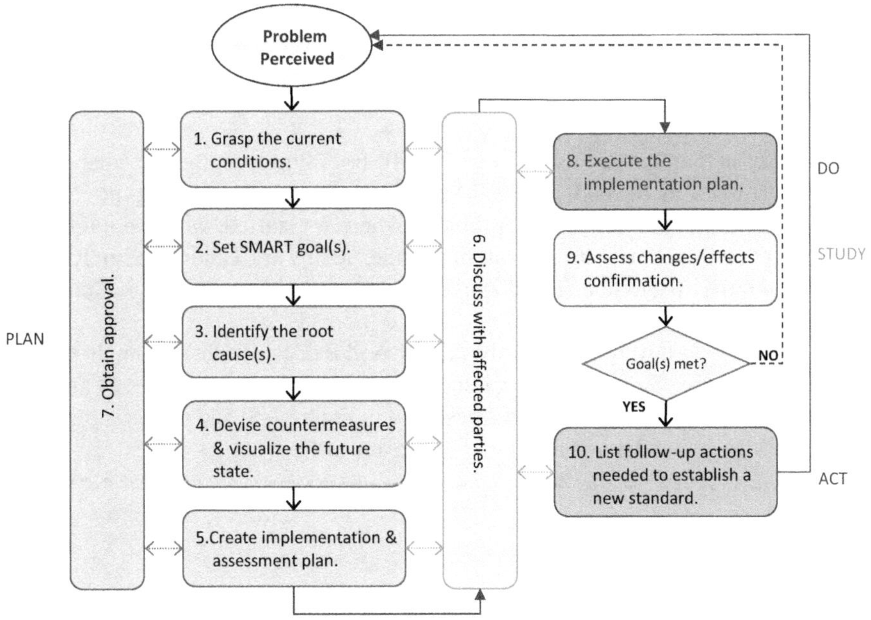

FIGURE 2.3 Revised A3 thinking process to match RIE briefs.

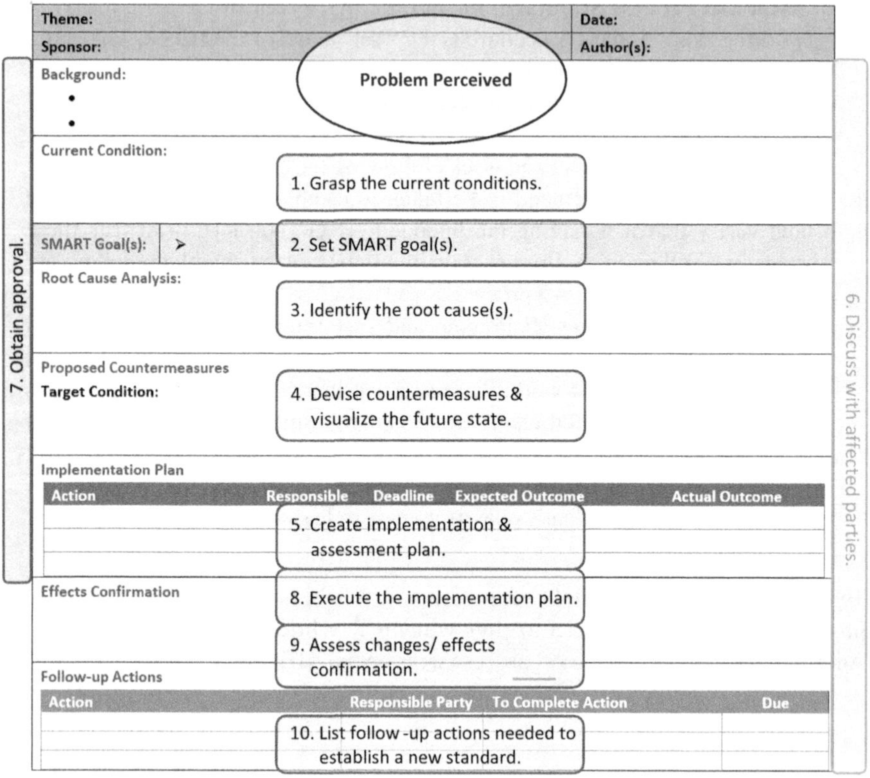

FIGURE 2.4 Guide on how to write and read RIE briefs following the A3 thinking process.

You may notice that each section of the RIE brief builds on the previous one in a logical narrative. By the end, anyone familiar with the context reading the RIE brief should understand the story of the problem: where we started, what we found, what we did about it, and how it turned out. In practice, teams draft and refine an RIE brief through iterative discussion. The document becomes a living summary of collective thinking and decision-making.

Now that we've covered our A3-thinking-based RIE briefs, let's bring them to life with a case study. We'll revisit the issue of delayed first-round surgeries and walk through an example of an RIE brief (presented in Figure 2.5) and discuss how a team utilized PDSA and A3 to achieve rapid improvement.

Case Study: Late First-round Surgery Operating Room (OR) Arrival Rates	Date: XX/XX/20XX
Sponsor: Operating Room Department	**Authors: Arlene and Bret**

Background

- Patients arriving at the Operating Rooms (ORs) behind schedule cause later procedures to have compounding delays.
- First-round patients are expected to be in the OR by 7:15, but historical data shows that most arrive between 7:25 and 7:35 on average.
- The cost of running the OR is between \$36 and \$62 per minute.

Current Condition

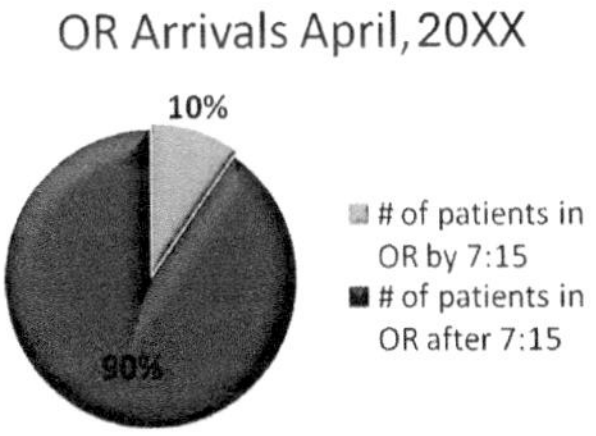

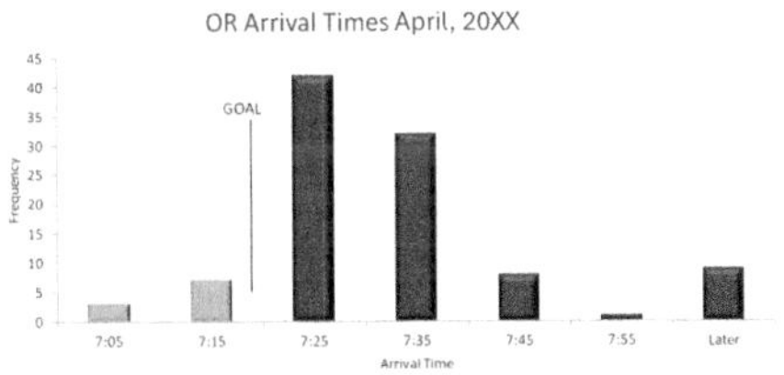

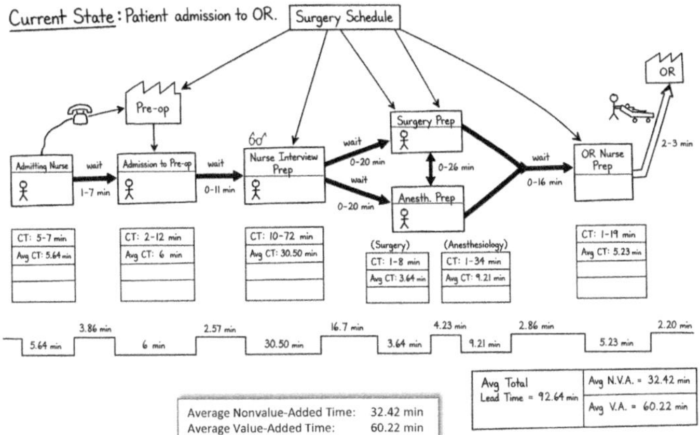

Average Nonvalue-Added Time:	32.42 min
Average Value-Added Time:	60.22 min
Average Total Lead Time:	**92.64 min**
Average Time Patient Arrives Before Surgery:	*90.00 min*

SMART Goal:	• **Reduce the average percentage of late first-round OR arrivals from 90% to less than 35%.**

Root Cause Analysis

Problem: Patients arriving late to the OR.

↳ **Why?** Anesthesiologist or Surgeon interview ends too close to 7:15 a.m.
  ↳ **Why?** The surgeon arrives at periop too close to 7:15.
    ↳ **Why?** The surgeon is unaware of the 7:15 goal.
      ↳ **Why?** The surgeon is unaware of what needs to happen after them.
        ↳ **Why?** There is no schedule designed for periop.

↳ **Why?** One Dr. waits for the other one to finish before coming into the room.
  ↳ **Why?** The surgeon and anesthesiologist arrived around the same time.
    ↳ **Why?** Insufficient communication/coordination
      ↳ **Why?** No established sequence.

↳ **Why?** Block **injection ended late.**
↳ **Why?** Block injection (BI) had delayed start.
↳ **Why?** The anesthesiologist arrived too close to 7:15 a.m.
↳ **Why?** BI equipment unavailable.
↳ **Why?** More than two BIs were scheduled at the same time.
↳ **Why?** Block injection took longer than usual.
↳ **Why?** Intravenous (IV) bag complications.
↳ **Why?** Patient concerns.
↳ **Why?** OR nurse prep ended after 7:15 a.m.
↳ **Why?** The OR nurse left the OR too close to 7:15.
↳ **Why?** The OR room was not ready by 7:15.
↳ **Why?** No standard procedure.
↳ **Why?** Staff can't find equipment.
↳ **Why?** Not labeled or no one accountable for equipment locations.
↳ **Why?** OR periop interview is long.
↳ **Why?** Patient questions.
↳ **Why?** Missing patient info.
↳ **Why?** Last minute doctor's orders.

Proposed Countermeasures

Target Condition:

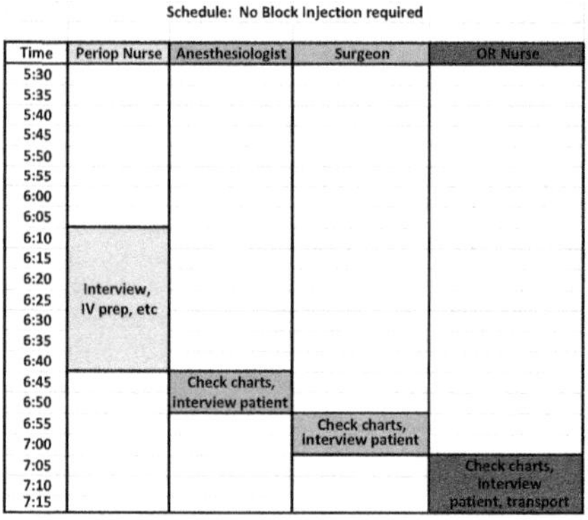

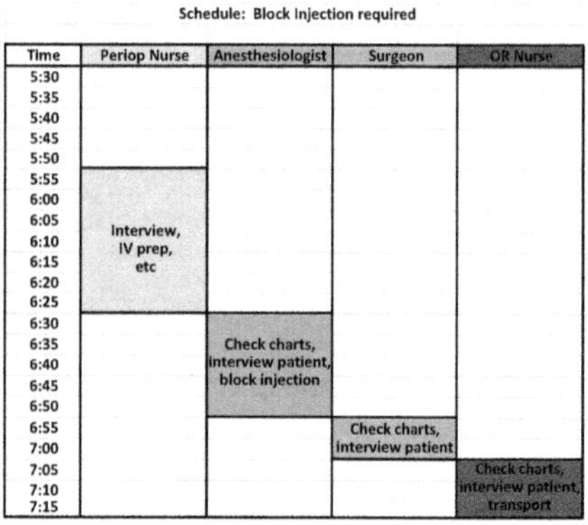

Implementation Plan

Action	Responsible	Deadline	Expected Outcome	Actual Outcome
1. Propose time schedules to OR and periop nurses.	Gert	Monday–Tuesday (week 2)	All nurses aware of the pilot plan.	Expected outcome met.
2. Propose time schedules to surgeons and anesthesiologists.	Gert	Monday–Tuesday (week 2)	All surgeons and anesth. aware of the pilot plan.	Expected outcome met.
3. Answer any staff questions.	Arlene and Bret	Monday–Tuesday (week 2)	Questions have been answered.	Expected outcome met.
4. Post copies of new schedules.	Gert	Monday–Tuesday (week 2)	Schedules posted in several key OR and periop areas.	Expected outcome met.

(Continued)

(Continued)

Action	Responsible	Deadline	Expected Outcome	Actual Outcome
5. Begin using proposed time schedules.	All nurses and doctors	Wednesday (week 2)	The pilot plan begins.	Expected outcome met.
6. Data collection of the pilot plan.	Bret and Arlene	Wednesday–Friday (week 2)	Data collected for all 7:30 surgeries.	Expected outcome met.
7. Present final report and follow-up actions.	Bret and Arlene	Friday (week 2)	The trial plan is either continued or not.	Trial took place.

Effects Confirmation

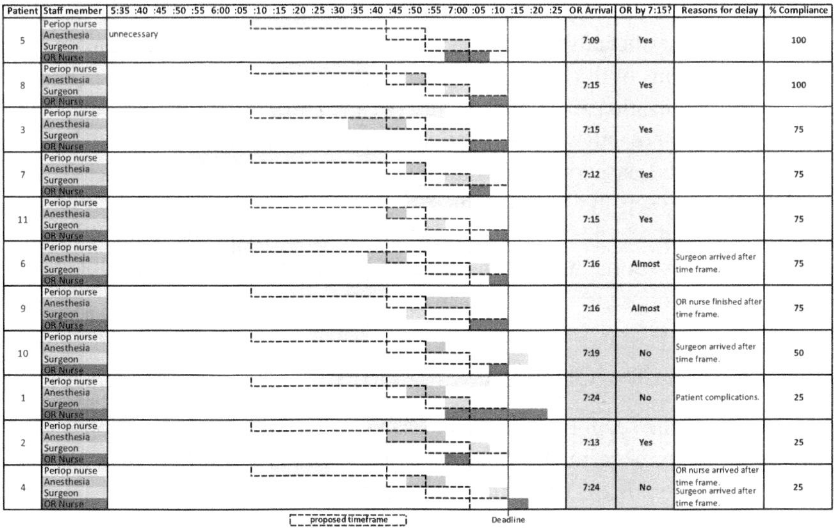

Results:

- Cases with ≥75% compliance (at least three out of the four stakeholders adhered to the schedule): **71% on-time.**
- Cases with <75% compliance (one or none of the four stakeholders adhered to the new schedule): **25% on-time.**

Follow-Up Actions

Action	Responsible Party	To Complete Action	Due
Continue the pilot study schedule for 1 or 2 weeks and confirm the pain block injection timeline. Then, assess the schedule and modify it as necessary.	Gert	Follow action.	In 2 weeks.
Continue to monitor compliance rates and give feedback to staff and doctors.	Gert	Follow action.	In 2 weeks.
Continue to work on getting buy-in from all staff members and doctors.	Gert	Follow action.	In 3 weeks.
Continue to identify additional sources of delay (e.g., lab tests, block injections, patient complications, etc.) and experiment with possible countermeasures.	Gert	Follow action.	In 4 weeks.

FIGURE 2.5 RIE brief on late first-round surgery OR arrival rates.

Background – The Problem and Why It Matters

Every morning, the OR schedule was being thrown off because first-case patients or providers were not ready at the scheduled time. Hospital policy expected the first patient to be in the OR by 7:15 a.m. for a 7:30 a.m. surgery. Still, most patients didn't arrive until 7:25 or even later. These delays had a domino effect: when the first surgery started late, subsequent surgeries ran behind schedule, staff had to work overtime, and patients' families experienced longer wait times.

The Background section also highlighted the financial stakes; running an OR costs roughly $36–$62 per minute, so even a 15- or 20-minute delay at the start of the day wastes thousands of dollars, strains the budget, and reduces the number of surgeries that could be completed.

The background established *why* fixing this problem was urgent: it was a source of operational waste, patient dissatisfaction, and staff stress. Additionally, the team framed the problem in terms of the hospital's goals. Punctual surgery starts are associated with better resource utilization and higher quality of care. By improving first-case start times, the hospital would not only save money but also potentially improve surgical throughput and staff morale (because a smooth-running OR makes everyone's day easier).

All this context sets the stage for the RIE. It tells anyone reading, "Here's the issue we're tackling and why it's important." As you read the RIE briefs, the Background section should answer the following questions, "What problem are we trying to solve, and why did we choose it?"

Current Condition – Quantifying the Delay

In the Current State section, the team described how things were functioning *before* any improvements were made. They gathered data on recent first-case start times and visualized the process of a patient getting to the OR. A simple chart showed the distribution of actual arrival times for first-case patients, illustrating that 90% arrived well past the 7:15 target. On average, patients arrived 10–20 minutes late (e.g., around 7:30), and as a result, first incisions were delayed accordingly.

Another aspect of the current state involved listing the key players involved: the patients (and their transportation from the ward or admission area), the perioperative nurses, the anesthesia team (often performing a preoperative nerve block), and the surgeons. All four groups needed to coordinate perfectly to hit that 7:15 ready mark.

To make the problem concrete, the RIE brief included a few baseline metrics: **~90% of first surgeries started late, with an average delay of around 15 minutes**. This meant the hospital was losing nearly 90 minutes of OR time per day (across six ORs), which, using the cost figures, resulted in hundreds, if not thousands, of dollars in lost productivity daily.

With the data collected and analyzed and process visualized, the first part of the RIE brief was reviewed with representatives from each stakeholder group – OR manager, perioperative charge nurse, surgical nurses and techs, lead anesthesiologist, and one of the surgeons who was deeply invested in resolving this issue. Everyone knew that it was a problem, but upon seeing the data, they were surprised to find the extent of the problem and were motivated to engage in the process to develop better procedures.

By laying out the current condition in this way and reviewing it with key stakeholders, the team ensured that everyone understood the reality of the situation and had the numbers to prove it. This also sets a baseline against which they could measure improvement. In summary, the Current State section answered, "What is happening now, and how bad is it?" It painted a quantitative picture of the problem that would later be used to evaluate whether the changes made a difference.

SMART Goal

The team's SMART goal was: "By the end of the RIE, we will reduce the percentage of late first-round OR starts from 90% to less than 35%." In other words, they aimed to reverse the trend, where almost all first cases were late, to one where the majority were on time. They chose 35% late as a realistic interim target – ambitious but attainable given the short timeframe of an RIE. In numeric terms, with roughly 10 first-case surgeries per week, the aim was to reduce the number of surgeries starting late from 9 out of 10 to just 3 out of 10.

Root Cause Analysis – Digging into "Why"

With the current state data in hand, the team next asked the all-important question: *why are these first-case surgeries starting late?* In the Root Cause Analysis section, they listed the main contributing factors they had discovered. Using a 5 Whys approach, they began with the obvious problem ("patients arriving at the OR behind schedule") and continued to probe for causes.

After these deliberations, the team summarized the root causes in the report. The major causes boiled down to two categories: communication failures (a lack of a standard procedure for notifying and preparing all parties for the first-case start) and schedule misalignment (tasks such as anesthesia blocks or patient transport weren't timed properly to meet the 7:15 ready time). In other words, there was a lack of standard work for the first-case morning routine. Everyone had the same goal (start on time), but there was no coordinated system to ensure it happened. By identifying these root causes, the team could now target their solutions to things that would truly make a difference, rather than just slapping a bandage on the symptoms.

Countermeasures and Target Condition – What "Good" Looks Like

The target condition envisioned *how* the morning process should ideally work after improvements. The brief described a future scenario in which all first-round patients, staff, and surgeons are ready by 7:15 a.m., with a reliable communication protocol in place. This implied that if the target condition were met, the nursing preparation would be completed first, followed by the anesthesiology team's completion of any necessary blocks. The surgeon would be present for an interview before surgery, and the OR nurse would have transported the patient into the OR by 7:15 a.m.

Based on the identified root causes, the team created two schedules: one for patients who require block injections and another for those who do not. The primary objective was to establish a standardized morning schedule for all stakeholders. The team created a simple timetable that back-planned from the 7:15 goal. For example, the perioperative nurse is ready by 6:10 (regardless of whether a block injection is needed), the surgeon finishes interviewing the patient by 7:00, and the patient is transported to the OR by 7:15.

The proposed schedules were then discussed with the key stakeholders to solicit input on their feasibility and ideas on how to make them better. The RIE facilitator also asked for their agreement to give the schedules a try for a few days to see how they work. Once agreement was reached, the schedules were clearly communicated to everyone involved with testing them, including perioperative nurses, OR nurses, anesthesiologists, and surgeons. The idea was to set a common expectation and make the timing explicit. This directly tackled the lack of a standard procedure.

It's worth noting that the chosen countermeasures did not require any major capital or new hires. They were mostly process or timing changes (and a bit of training/communication effort). This is typical of many RIE countermeasures. They often involve changing workflows or information flows, reorganizing workspaces, or providing signage or visual instructions rather than capital investments. In this case, for example, rather than adding more staff to rush patients to the OR (which some might have suggested initially), the countermeasure was better coordination of the perioperative and surgical teams.

Implementation Plan – Making It Happen

With the countermeasures decided, the team moved to the Implementation Plan. This was essentially a to-do list outlining who would do what and by when to implement the countermeasures. The plan was laid out in a table, with columns for Action, Responsible Person, Timeline, and Expected Outcome. This implementation plan was basically a mini project management chart that ensured nothing important fell through the cracks. Each task had an owner and a deadline, which creates accountability. By writing this out in the brief, anyone reviewing the plan could immediately see if the timeline made sense and if the responsibilities were clear. For instance, leadership could ask, "Have the anesthesiologists agreed to this change?" and the team could point out that they were consulted in Action 2 and involved in the pilot.

Once again, the team reviewed the implementation plan with all actors to get their commitment to taking the action by the deadline indicated and buy-in to test the new coordination schedule to see if it works. This also gave the stakeholders opportunity to request adjustments to the implementation plan if needed.

Effects Confirmation

The data showed a dramatic improvement. On the days of the pilot, whenever most stakeholders followed the new process, first-case on-time starts increased dramatically. Specifically, in cases where at least three out of the four key players adhered to the new schedule, 71.43% of those surgeries started on time. On the other hand, when two or less of those players followed the timeline (meaning poor compliance), on-time starts were a dismal 25%. This comparison vividly demonstrated the value of the coordinated approach. When most of the team did their part, they beat the target (they got it to roughly 30% late, which is within the <35% goal). When the plan wasn't followed, the old problems persisted.

Goals Met?

A good practice for RIE briefs is to state whether the goals were met. It can be seen from the report that the team stated this by using a check mark and quantifying the improvement in bold:

- Cases with ≥75% compliance (at least three out of the four stakeholders adhered to the schedule): **71% on-time**.

Follow-Up Plan to Revisit the System

The team didn't declare victory and walk away. They used the last section to outline the actions necessary to move from the pilot study to a standard procedure for all cases and personnel. From the snippet in the case document, we see that they proposed continuing the pilot for another 1–2 weeks to confirm the improvement and tweak details, such as the nerve block timing. They also planned to keep monitoring compliance rates and give feedback to staff and physicians. For example, if certain surgeons are not consistent on time, the OR director would have one-on-one conversations (a gentle enforcement of accountability). The follow-up actions essentially said: "Here's how we will continue to check and adjust." It included who was responsible and target dates (e.g., "Gert to review compliance data in two weeks").

Another follow-up step was about buy-in. The team recognized that sustaining the change required everyone's commitment, so they planned to continue communicating and working on achieving 100% staff buy-in. Sometimes that meant addressing concerns ("This new schedule means I have to come 10 minutes earlier; is that paid?") or simply reinforcing positive results ("Look, we did great last week, let's keep it up!"). They recorded this intention in the report so it would not be forgotten.

Ultimately, the case study's A3 story had a successful outcome. The OR consistently improved its first-case start performance over the following weeks, and the culture began to shift, where an on-time start became the expected norm rather than a pleasant surprise. By documenting the entire journey in an RIE brief, the team could easily share the results with hospital leadership and even other departments.

This case study exemplifies how improvement in healthcare can be achieved through the lens of A3 thinking. We saw how a pressing problem was addressed not by guesswork or argumentation among strong personalities, but through a structured method grounded in data and facts about how things actually run: defining the problem, analyzing it deeply, setting a clear goal, implementing targeted fixes, and measuring the impact. The A3 process guided the team through logical steps and kept their efforts aligned with the real issues. The RIE brief provided a communication tool to get everyone on board and later to showcase the improvement.

You, as a reader and future practitioner, can apply the same approach to a myriad of healthcare challenges, whether it is reducing clinic wait times, improving medication safety, or streamlining hospital admissions. The specific countermeasures will differ, but the A3 thinking process and structure will ensure you're systematic and effective in whatever problem-solving adventure you embark on.

REFERENCES

Agency for Healthcare Research and Quality (AHRQ). (2015). *Using root cause analysis to improve quality and performance*. Retrieved February 2025, from https://www.ahrq.gov/evidencenow/tools/root-cause-analysis.html

Kumah, A., Nwogu, C. N., Issah, A.-R., Obot, E., Kanamitie, D. T., Sifa, J. S., & Aidoo, L. A. (2024). Cause-and-effect (fishbone) diagram: A tool for generating and organizing quality improvement ideas. *Global Journal on Quality and Safety in Healthcare*, 7(2), 85–87. https://doi.org/10.36401/JQSH-23-42

Sobek, D. K., & Smalley, A. (2011). *Understanding A3 thinking: A critical component of Toyota's PDCA management system*. CRC Press. https://doi.org/10.4324/9781439814055

3 A Guide to Rapid Improvement Events in Healthcare

Rapid Improvement Events (RIEs), also known as "Kaizen events", are used in healthcare to drive short, focused improvement. *Kaizen* is a Japanese term for a philosophy of continual improvement, in which small, incremental changes made over time lead to significant performance improvements. An RIE brings together a multidisciplinary team to analyze a specific process or study a specific problem or opportunity, implement changes, and achieve quick, tangible performance improvements.

RIEs are valued for producing results in days, not months, by eliminating waste and streamlining processes in ways that improve patient experience and reduce staff frustration (most of the time, with a low budget). These events are highly structured and time-boxed, typically running 5–10 consecutive days of intensive work. During this period, participants map the current process, identify problems, implement solutions on the spot, and standardize the new process. RIEs are characterized by immediate implementation of improvements during the event, rather than just planning work (U.S. Environmental Protection Agency, 2017).

Normally, a facilitator leads a cross-functional team in defining the problem, observing processes, collecting and analyzing data, generating ideas, implementing changes, and measuring results. Successful RIEs result in some implemented changes but often require post-event actions to implement longer lead-time changes and ensure that the improvements are sustained; thus, follow-up action plans are also a typical deliverable.

STAKEHOLDER ENGAGEMENT AND TEAM COMPOSITION

A defining feature of RIEs is the engagement of frontline stakeholders in the improvement process, that is to say, the people who actually perform the work. Organizational improvement professionals have learned that improvements are most effective and sustainable when they are designed and implemented by the people who do the work, with support from leadership, rather than imposed top-down (NHS England, n.d.). Thus, building the right RIE team and fostering stakeholder buy-in are critical to sustaining change. The following are the key factors to consider when forming a team.

- **Multidisciplinary Team**: RIE teams in healthcare are typically cross-functional, bringing together staff from all areas that touch the targeted process or operation. For example, an RIE to streamline surgical instrument sterilization might include sterile processing staff, operating room nurses or technicians, surgeons, anesthesiologists, quality/safety personnel, and one or more hospital administrator overseeing these areas. Including

DOI: 10.1201/9781003670698-3

representatives of each stakeholder group ensures that all perspectives are heard and that proposed solutions consider the realities of each role.

- **Clinician and Support Staff Involvement**: In clinical process improvements, physician participation is often crucial. RIE case studies have noted that involving doctors can be challenging due to their busy schedules, but it pays dividends in terms of smoother implementation. Similarly, including relevant support departments, such as IT or informatics staff, is recommended given the reliance on electronic health record systems and other technologies in most processes (AHRQ, 2014). If the process crosses departmental boundaries (as many do in healthcare), involving representatives from those departments avoids a short-sighted approach. Broad engagement during the RIE means the team can address interdepartmental handoffs and data systems issues in real time, rather than discovering obstacles after the event.
- **Frontline Empowerment and Culture**: The philosophy of an RIE is that the *people who do the work are the ones best equipped to improve it.* Leaders from various health systems observe that RIEs create a sense of ownership, as staff see that their expertise is valued and their ideas drive change (AHRQ, 2014; Nino et al., 2020). In contrast to traditional top-down initiatives, RIEs give frontline caregivers a voice in designing solutions, which can greatly increase buy-in and morale. They will work hard to ensure a solution works and is sustained because, after all, it was their idea. In fact, on many occasions, when those stakeholders return to their roles after the event, they often come up with ideas on how to further improve their processes (Nino et al., 2020). However, to harness this effect, the organization must create a psychologically safe environment for staff to speak up and contribute openly (AHRQ, 2014). Effective RIE facilitation fosters trust, encourages all voices to be heard, and avoids hierarchy, ensuring that a nurse, a doctor, and a clerk on the team all have an equal say throughout the event.
- **Leadership and Sponsorship**: Stakeholder engagement also extends to leadership and sponsorship. Each RIE team should have an executive sponsor or process owner who provides direction and can swiftly address escalated issues, e.g., approving an experiment or reallocating resources (AHRQ, 2014). The sponsor typically helps the team align improvements with organizational goals and ensures barriers are removed. That said, during the event itself, leaders are usually asked to "get out of the way" of the team's creative process. Their role is to support the team, not micromanage it.

Engaging the right stakeholders in an RIE is vital for both practical and cultural reasons. A well-composed team with the necessary process knowledge, the authority to implement changes, and diverse perspectives will think creatively and devise more robust solutions.

RIE DURATION

The classic RIE format is often a week-long (5-day) event, which has become a common model in health systems like the NHS and Virginia Mason Medical Center

(Meguerdichian et al., 2017). While this is the standard that has been adopted, variations in length do exist. Some experts note that RIEs can range from short 1–2 day blitzes up to multi-week events, depending on the scope and complexity of the problem (U.S. EPA, 2017). Regardless of length, the core structure of an RIE remains consistent: thorough planning and scoping are done beforehand, followed by an intensive, collaborative session where the problem and work are studied, changes are designed and tested in real time, and rigorous follow-up is conducted to sustain the gains (Haapatalo et al., 2023).

Successful RIEs are built on extensive upfront preparation. Studies and change management guides alike emphasize that RIE success is directly linked to how well the event is planned. Key preparation steps typically occur in the weeks (or even months) before the event and include:

- **Defining the Problem and Scope**: A clear problem statement, project scope, and goals should be agreed upon and documented, typically in a project charter or Statement of Work, well before the RIE.
- **Securing Leadership Support**: High-level leadership should not only approve the RIE project but also align it with strategic objectives and commit necessary resources. Executive sponsors play a vital role in removing barriers and empowering staff to implement changes (Toussaint & Berry, 2013).
- **Forming the Team and Engaging Stakeholders**: The right mix of people must be invited to the event. Planners should identify all key stakeholders of the process, including frontline staff who perform the work, managers, physicians, or other clinicians, and support departments such as IT, and occasionally patients or external partners.
- **Securing Time from the Participants**: Ideally, team members should be freed from their normal duties for the event so they can focus fully (AHRQ, 2014). In healthcare, backfilling staff shifts or adjusting schedules is often necessary to allow doctors, nurses, or support staff to take time off their regular duties and participate.
- **Gathering Baseline Data**: It is essential to collect metrics on current process performance before the RIE. Baseline measurements allow the team to quantify improvements achieved and set targets. Many RIE guides recommend identifying a few key performance indicators and collecting data for a period preceding the event.

The case studies presented in this book were all part of 2-week RIEs because, in most cases, the facilitator needed to do some of the preparation work (such as data collection) that would normally occur before a 5-day RIE. That preparation work, mostly involving data collection and analysis, happened during the first week with the second week focused on brainstorming, developing, and piloting countermeasures. Two-week RIEs are also useful when a project's scope requires more in-depth analysis, experimentation, or stakeholder training than a typical week-long event can accommodate.

THE TWO-WEEK RIE

This book is based on a decade of experience running a healthcare engineering summer internship program. Industrial Engineering students, equipped with an array of systems engineering and analysis skills, were placed at clinics and hospitals to facilitate RIEs under the guidance of a coach (typically one of the authors). Hospitals and clinics submitted proposals to participate in the summer program, including two potential projects and guarantees of sponsorship.

Four to eight weeks prior to the facilitator's arrival at the facility, we worked with the hospital's management to flesh out, in broad terms, the problem or issue they were facing, define the main objectives of the improvement project, and identify the metric(s) they were most interested in improving. The project sponsor(s) and stakeholders were identified by name. The project scope was delineated, and any existing data were identified that would be useful to the team's effort. After that meeting (sometimes it took more than one meeting), the project sponsor was responsible for informing the stakeholders about the project, soliciting their cooperation, and making room in their schedules to participate prior to the event. They also collected the data identified to have it ready for the start of the RIE.

With this preparation work completed, the typical two-week (ten working days) RIE agenda proceeds according to the following timeline:

Day 1	The RIE facilitator meets with key contacts, is oriented to the facility and specific work area of concern, ascertains the desired outcomes of the project, does a Gemba walk, conducts some initial observational studies, and drafts a Statement of Work. The facilitator uses the pre-visit work as a starting point in defining the RIE scope and objectives, but modifies and extends them based on information gathered on the ground. The Statement of Work is reviewed by the project sponsor and other key administrators at the facility to ensure a well-defined and appropriate scope, clear and measurable goals, and a realistic and appropriate plan for the RIE.
Days 2–4	The facilitator collects data related to the project objective, including archival data that the host facility may already have on file, as well as observation and documentation of existing processes, patient or flow volumes, processing times, and any other necessary data to understand the current system. Data are documented and analyzed; stakeholders review methods and results.
Days 4–5	The facilitator ascertains the root causes of the problem and verifies them with the parties involved. Additional data collection may be required. Idea generation often begins.
Days 5–6	The facilitator compiles ideas to address the root causes by brainstorming and soliciting ideas from hospital or clinic staff. Ideas are evaluated, and collectively all affected staff members and administrators decide which changes to implement.
Day 7	The facilitator creates an implementation plan that specifies actions, deadlines, owners, and expected outcomes for each, and solicits agreement from the parties involved.
Days 8–9	Changes are implemented. The facilitator measures the level of improvement; visual tools are preferred.
Day 10	The facilitator incorporates the results into a final report and meets with management to conduct a debriefing.

A common approach is for the RIE facilitator or coach to remain engaged with the team for some time after the event. For example, at one hospital, after an RIE was completed, the facilitator created a list of action items that needed follow-up. In the first 30 days, the facilitator met with the staff daily for 15–20 minutes to check progress on the action plan. In the next 60 days, meetings were held weekly as the process changes solidified (AHRQ, 2014). These check-ins enabled the team to address new issues, ensure that new standard procedures were being followed, and make any necessary adjustments.

In our RIE briefs, we use a table format to capture and organize post-event actions. This approach clarifies responsibilities by assigning each action item to an owner with a specific timeline to encourage continuation of improvement beyond the event's conclusion. Many of the follow-up tasks naturally generate opportunities for additional RIEs, thereby extending the impact of the original project and encouraging teams to continue seeking the next area of improvement. By documenting these items systematically, we reinforce accountability, provide a record for future reference, and, most importantly, embed the idea that improvement is an ongoing process. In this way, the follow-up table serves as more than a checklist: it becomes a tool that helps sustain momentum and support a culture of continuous improvement.

Despite best efforts, not all RIE improvements will sustain at 100% over the long term. Healthcare is a complex and dynamic environment; staff turnover, fluctuating workloads, or downstream changes can erode the gains made. However, studies show certain factors improve the odds of sustainability. A supportive work culture and continuous improvement mindset are chief among them (Haapatalo et al., 2023). In a 2023 mixed-method study across 26 Kaizen events, units that sustained improvements had strong motivation for improvement and a culture that encouraged it, whereas units that struggled often cited "lack of time" and high daily workload as barriers (Haapatalo et al., 2023). This suggests that leadership must ensure staff have time for improvement activities and that recognition/incentives for sustaining improvement gains are in place. Making the RIE changes "stick" may involve incorporating new process adherence into manager evaluations or continuing to highlight the achieved improvements in staff meetings to prevent them from being forgotten. Figure 3.1 summarizes the 2-week RIE following the A3 thinking process in the RIE brief format discussed in the previous chapter.

It can be seen from the figure that approximately 70% of the time is dedicated to the Planning stage of the Plan–Do–Study–Act (PDSA) cycle. The Do and Study stages (executing the implementation plan and assessing changes) are typically completed over the last 2–3 days. While careful planning is essential before testing solutions, it is equally important to move forward with implementation and measure the results.

LESSONS LEARNED FROM OUR RIEs

Every RIE is different, shaped by its people, culture, and context. Yet, after guiding over eighty of these events, we've noticed themes that repeat themselves time and again. We share some of those lessons here.

1. **Most problems stem from communication issues**. Nearly every root cause analysis we did had broken communication as a key contributor to the issues or problems being experienced. Some of the more common issues include:

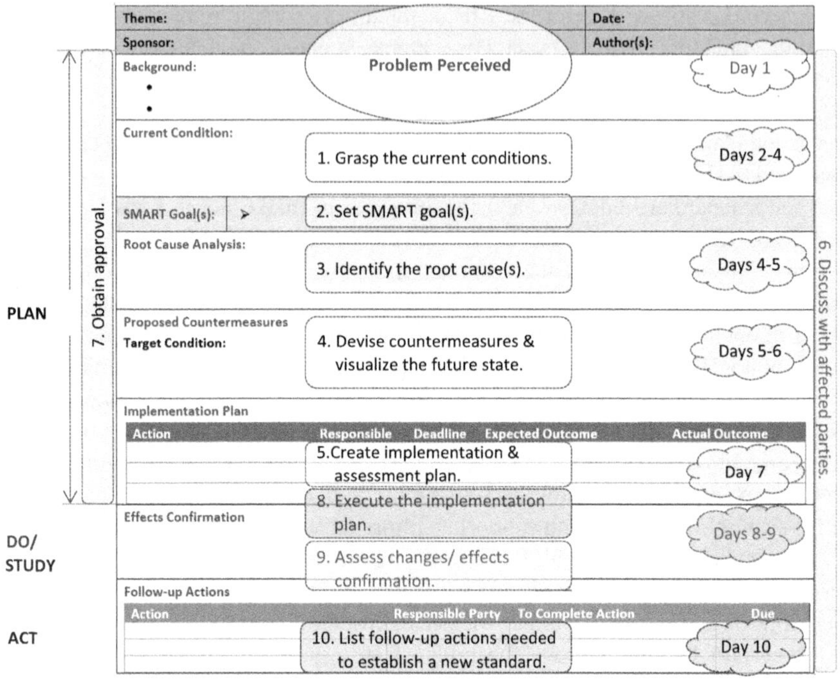

FIGURE 3.1 Guide for a 2-week RIE following the A3 thinking process.

- Breaks in the handoff chain (e.g., providers too busy with primary duties to transition patients properly).
- Misunderstandings, misalignment, or unspoken expectations.
- Unclear assignment of roles and responsibilities.
- Failure to communicate responsibilities and expectations effectively.

2. **Improve even with incomplete data.**
   - RIE teams often wish they had larger sample sizes, more accurate counts, or data from a longer time frame or larger range of patients. But our advice is as follows: don't wait for perfect information. Small improvements are better than no action at all.
   - Do the best with the data you have and then learn through experimentation.
3. **Do regular Gemba walks.**
   - Visit the workplace at least once or twice a day to see the process firsthand.
   - Interact with staff to build rapport and gather additional information and insights.
4. **Involve everyone affected and secure management support.**
   - Broad participation ensures the development of realistic and sustainable solutions.

   - Involving stakeholders and communicating with leaders at each stage of the process gives them confidence in the data and facilitator.
5. **Define scope and create a solid Statement of Work.**
   - Setting clear project boundaries and objectives at the start of event is crucial to success.
6. **Each problem is unique.**
   - Thoroughly understand the problem and the context first so you can devise specific countermeasures customized to that situation.
   - Standardization comes from the systematic method you use to understand the processes and issues before jumping to solutions.
7. **Emphasize implementing improvements, not just defining problems.**
   - Focus on taking action, not just describing what is wrong.
   - Framing implementation as an experiment can overcome initial resistance from stakeholders who may fear getting stuck with a process that doesn't work for them.
8. **Brainstorm freely when designing countermeasures.**
   - At the future state stage, treat it as open idea generation.
   - There are no impossible ideas. Instead of asking "how?", ask "why not?"
9. **Expect to refine during implementation.**
   - Most countermeasures will not work perfectly the first time; learn, adjust, and improve as you go.
10. **Leverage your network.**
    - There is strength in your network. As Porter Gale said, "Your network is your net worth." Build relationships with people who share similar ideas.

Scan the QR code below and join this book's LinkedIn group (https://www.linkedin.com/groups/13398009/).

There, we plan to discuss practical applications of PDSA and A3 thinking, share experiences from improvement projects, and connect with professionals who are passionate about making healthcare systems safer, more efficient, and more patient-centered.

LOOKING BACK, MOVING FORWARD

Now that we have walked through the two-week RIE format, along with the PDSA and A3 thinking processes, it is your turn to explore the 30 real case studies that make up the heart of this book. These stories illustrate how the principles we've discussed come to life in practice.

A few notes on our case studies:

1. For the 30 reports presented in this book:
   - We removed or replaced the actual date the study occurred and the location to protect the identity of the project sponsors.
   - All facilitators' and participants' names have been replaced with names of North Atlantic hurricanes from the last 15 years, provided by the United States National Oceanic and Atmospheric Administration (https://www.nhc.noaa.gov/-aboutnames.shtml).
   - The deadlines have been replaced with day of the two-week RIE, while the year or dates in relation to the RIE are hidden (i.e., 20XX). For example, something that was due on the Tuesday of the second week will be read as *Tuesday (wk.2)*.
2. With the previous three changes in mind, it is important to emphasize that the location, date, or individuals involved in the RIEs are not the focus. What matters here is the systematic approach to improving processes, identifying root causes, and developing effective solutions. The goal is to address systemic problems rather than assign blame to people.
3. Every SMART goal was formulated to be achievable and time-bound by the duration of the RIE. In other words, the achievable and time-bound components of the SMART goal(s) come from the premise: *By the end of the RIE, we will ...*
4. A specific issue had already been identified by the various hospital departments and clinics prior to initiation of the RIE, along with the scope of the project. The preliminary work focused on project scope and identifying the RIE team, but most of the data collection, validation, and analysis work occurred during the 2 weeks.
5. Because the RIEs were time-limited to two weeks, the team could not implement everything needed for full implementation of the countermeasures. Additionally, further improvements were often identified during the pilot of the countermeasures. Therefore, all RIE briefs conclude with a list of to-do items (in the Follow-up Actions section) for the sponsoring department to carry out over the following weeks to fully implement and sustain the changes.

Now, before we move forward, it is worth taking a step back to reflect on the goals we set at the beginning of this book. We're pleased to say we achieved one of our SMART goals and got really close to the second one: we have included exactly 30 RIE case studies and the opening three chapters account for just a little over 25%

of this book's content (according to our calculations, they make up about 25.35% of the book content before editing by the publisher). *The last two goals are now in your hands.*

At the start of this journey, we imagined being lost at sea, waiting for the sunrise to guide us east. Now, as you embark on your improvement journey, we wish you fair winds and following seas. May the tools of PDSA, A3, and RIEs be the sun on your horizon, giving you direction, clarity, and the courage to move forward with purpose.

REFERENCES

Agency for Healthcare Research and Quality. (2014). *Lean implementation case studies*. U.S. Department of Health & Human Services. https://www.ahrq.gov/research/findings/final-reports/leanimpl/index.html

Haapatalo, S., Holmström, S., Alenius, H., & Eloranta, S. (2023). Success in preparation and follow-up was associated with sustained results of Kaizen events in hospitals: A mixed-method study. *BMJ Open*, 13(8), e071743. https://doi.org/10.1136/bmjopen-2023-071743

Meguerdichian, M. J., Bajaj, K., & Falcone, R. A. (2017). The OR turnaround: A rapid cycle improvement project using simulation. *BMJ Simulation & Technology Enhanced Learning*, 3(2), 75–80. https://doi.org/10.1136/bmjstel-2016-000147

NHS England. (n.d.). *PDSA cycle template and learning guide*. NHS England. Retrieved August 3, 2025, from https://www.england.nhs.uk/improvement-hub/wp-content/uploads/sites/44/2015/08/learning-handbook-pdsa.pdf

Nino, L., Claudio, D., Valladares, L., & Harris, S. (2020). An enhanced Kaizen event in a sterile processing department of a rural hospital: A case study. *International Journal of Environmental Research and Public Health*, 17(23), 8748. https://doi.org/10.3390/ijerph17238748

Toussaint, J. S., & Berry, L. L. (2013). The promise of Lean in health care. *Mayo Clinic Proceedings*, 88(1), 74–82. https://doi.org/10.1016/j.mayocp.2012.07.025

U.S. Environmental Protection Agency. (2017). Lean in government starter kit, Version 4.0. https://www.epa.gov/sites/default/files/2017-11/documents/lean-starter-kit-version-4.pdf

4 Supply Management Case Studies

The first set of Rapid Improvement Event (RIE) briefs presented deals with the ubiquitous challenge facing all healthcare facilities: medical supplies management. As you review the RIE briefs, you will note that although the goals for the projects are similar, each situation and context is different, and therefore, the solutions proposed and adopted are different. You will not find a one-size-fits-all set of solutions, although you will recognize some common themes associated with effective supply management, notably: logical organization of the physical inventory, visual management, and simple tracking and reordering processes.

Case 4.1 addresses the management of medical supplies on the medical floor and the lost charges of supplies used. Within the two weeks, the team implemented a clear system for charge capture and implemented an inventory management system, resulting in an immediate 10% reduction in excess supplies.

Case 4.2 dealt with the operating room (OR) supply organization and management. The RIE resulted in clearer and more usable organization of supplies within the OR and 15% reduction in search time to locate supplies. Additional opportunities for further improvement were identified in the pilot run of the new supply locations and labeling.

The RIE documented in Case 4.3 devised an improved supply management system for an emergency department (ED). The team achieved a 50% reduction in the time to locate supplies, and a 50% reduction in the number of items stored in the ED, without increasing the risk of stockouts.

The team in Case 4.4 tackled the central supply room at a small hospital. Through a small number of inexpensive and no-cost changes, they significantly enhanced the ability of hospital personnel to locate and replenish needed supplies elsewhere in the facility. They realized a 57% reduction in distance traveled within the central supply room and a commensurate 71% reduction in time to locate supplies.

Finally, Case 4.5 addressed a post-COVID challenge of dealing with an excess amount of emergency supplies. The heavy influx of supplies during the peak of the COVID pandemic resulted in the hospital having more inventory than it could handle, and without an accurate idea of how much inventory it had. The project determined the amount of inventory and then developed a sustainable system for managing emergency supplies, balancing cost with emergency preparedness by centralizing and right-sizing their emergency supply inventory. The result was a 36% reduction in inventory.

DOI: 10.1201/9781003670698-4

While the cases are all different, each team followed the same high-level collaborative problem-solving approach to achieve substantial improvements. You will notice charts, diagrams, and images that were used to communicate and drive consensus around the key issues. You will see that multiple alternative ways to address the issues (i.e., countermeasures) were presented and evaluated. You will gain insight into the detailed implementation plans that allowed for efficient and effective project implementation. You will also see how the facilitators used data and quantitative metrics to inform their ideation and demonstrate the effectiveness of the countermeasures.

Case Study 4.1: Improving Inventory Ordering System and Billing Process	**Date: XX/XX/20XX**
Sponsor: Nursing Department	**Author: Alberto**

Background

- 22 bed Critical Access Hospital that serves many long-term care patients.
- 6 registered nurses (RNs), 8 certified nurse assistants (CNAs), 4–6 traveling nurses.
- No inventory management system.
- No standard system to issue charges for supplies inpatients use.
- Time nursing spends managing inventory takes away from time available for patient care.

Current Condition

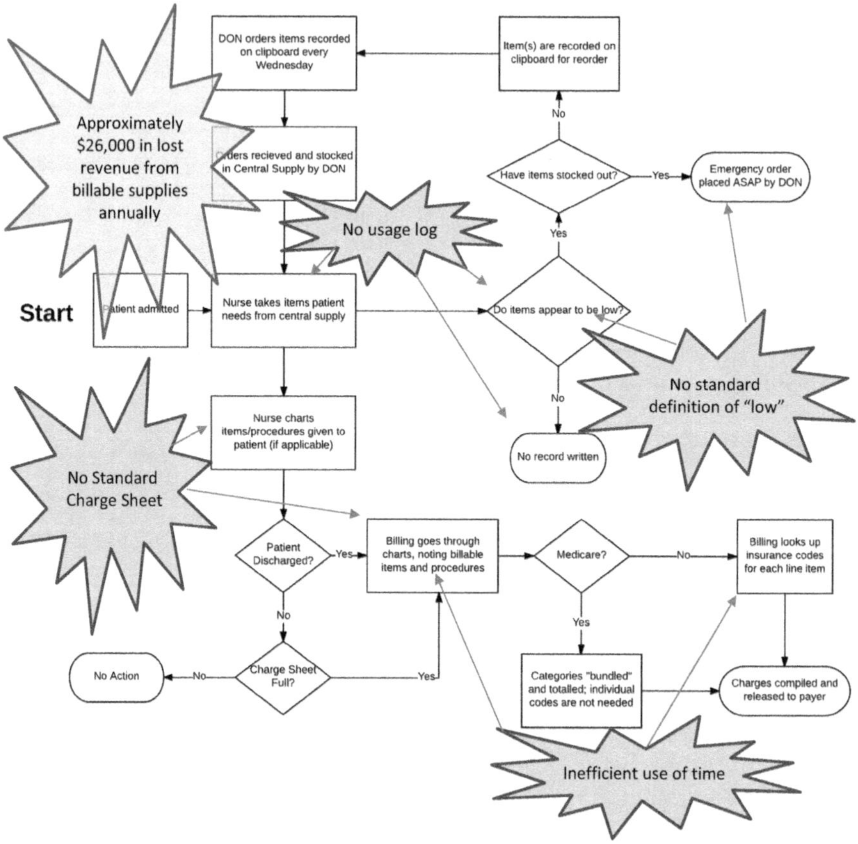

Lead time for supplies is estimated to be 7 days.
Central supply contains approximately 200 stock keeping units (SKUs) and 9,000 individual items.

SMART Goals:	• **Create a standardized inventory management process and Charge Sheet.** • **Create a Charge Sheet to increase revenue from billable med/surg supplies by 100%.** • **Reduce current excess inventory by 5%.**

Root Cause Analysis

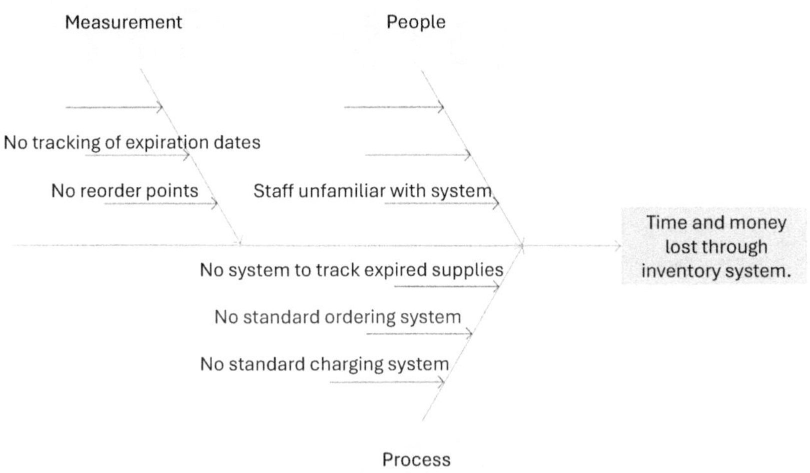

Proposed Countermeasures

- Par-level inventory system.
  - In general: T = 1 week, LT = 1 week, SS = 4 weeks for high usage items, 1 week for low usage items.
- Charge sheet for inpatients and possibly Emergency Room (ER).

Inventory Variations:

1. **Periodic Review**: Inventory central supply on weekly/monthly basis; order up to par level designated in Excel spreadsheet.
2. **Daily Ordering**: Kanban card placed at reorder point; order fixed quantity, calculated via spreadsheet, to return to par.
3. **Continuous Tracking**: Sign out supplies for each use; orders at reorder point up to max level; Master database in Microsoft Access to track orders and usage.

	Pros	Cons
1.	• Regular order placement. • Amount to order is easily determined. • Very easy to implement.	• Must take inventory of supplies on a regular basis. • No visual indication of when to reorder. • Additional stock-out risk if demand surges. • Requires more safety stock than other methods.

(*Continued*)

(*Continued*)

	Pros	Cons
2.	• Visual cue when supplies need reordered. • Fixed quantity of supply ordered each time. • Responsive to change in demand. • Simple to add items and change pars.	• Supplies should be ordered as soon as card is received to decrease the risk of stock out. • Almost 200 cards; risk of losing, cost of production/replacement.
3.	• Quick and easy to see expirations; what needs reordered. • Log could be accessed by Chris as a charge sheet. • Order placement period is flexible. • Simple to add items and change pars. • Responsive to changes in demand.	• Labor intensive: depends heavily on nurse participation to sign out supplies, which must be logged into a database file by DON. • Complex to implement and upkeep. • No visual cue in the central supply room.

Kanban system (Variation 2) chosen by stakeholders.

Implementation Plan

Action	Responsible	Deadline	Expected Outcome	Actual Outcome
Train RNs and CNAs on new systems	Alberto/ Beryl	Ongoing	Explain new ordering and charge system.	About 2/3 of staff trained; those not here this week will be brought up to speed by Beryl next week.
Gather materials for Kanban	Beryl	Tuesday p.m. (week 2)	Colored cardstock purchased; laminator procured.	Cardstock and laminator available for use Tuesday a.m.
Develop spreadsheet to calculate par levels and reorder points	Alberto	Tuesday p.m. (week 2)	Par levels and reorder points calculated.	Draft done Friday p.m.; finalized Tuesday a.m.
Finalize Emergency Room (ER) charge sheet	Chris/Beryl/ Alberto	Tuesday p.m. (week 2)	ER charge sheet and acuity sheet.	Meeting got moved back; finalized Wednesday p.m.
Manufacture Kanban cards	Alberto	Wednesday a.m. (week 2)	Cards laminated; appropriate fields written in.	Finished Tuesday p.m.; location needs to be determined for two cards.

(*Continued*)

(*Continued*)

Action	Responsible	Deadline	Expected Outcome	Actual Outcome
Post Kanban bin to receive cards	Beryl/ Alberto	Wednesday p.m. (week 2)	Receiving bin posted in DON office.	One bin posted in each storage area; one bin in DON office by Tuesday p.m.
Place Kanban cards	Alberto	Wednesday p.m. (week 2)	All 200 Kanban cards posted.	All 200 cards posted Wednesday a.m.
Add expired tracking to spreadsheet	Alberto	Thursday a.m. (week 2)	Develop way to track expired items within inventory spreadsheet.	Draft done Tuesday AM; nursing decided to go a different direction, so will not be finalized.
Finish rough draft of charge sheet	Alberto	Thursday p.m. (week 2)	Charge sheet template completed.	Inpatient sheet done and edited; decided to create an ER sheet as well.
Design Kanban cards	Beryl/ Alberto	Friday a.m. (week 2)	Determine information to be contained on card; color; layout.	Purple chosen as card color; information to be contained finalized by noon Friday.
Finalize inpatient charge sheet	Alberto/ Chris/ Beryl	Saturday a.m. (week 2)	Charge sheet prepared to be deployed on floor.	Finished Thursday afternoon for inpatient sheet; ER sheet still in progress.
Evaluate organization/ location of supply storage	Alberto	If time	Evaluate signage, organization, and placement of supplies within storage spaces and within the hospital.	Shelves signed; minor reorganization within shelves, but no large-scale reorganization of shelves.

Effects Confirmation

Description of New Standard Process

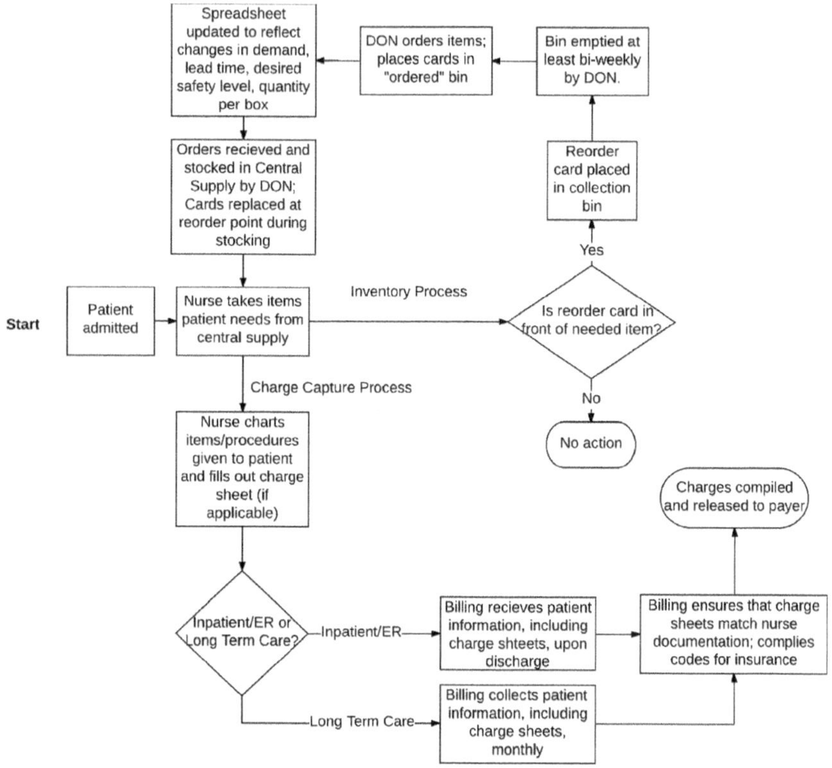

Inpatient Charge Sheet

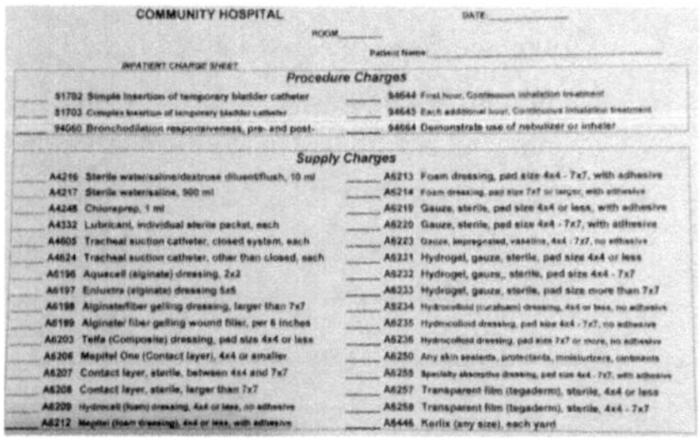
COMMUNITY HOSPITAL DATE ________

ROOM ________

Patient Name ________________

INPATIENT CHARGE SHEET

Procedure Charges

____ 51702 Simple insertion of temporary bladder catheter	____ 94644 First hour, Continuous inhalation treatment
____ 51703 Complex insertion of temporary bladder catheter	____ 94645 Each additional hour, Continuous inhalation treatment
____ 94060 Bronchodilation responsiveness, pre- and post-	____ 94664 Demonstrate use of nebulizer or inhaler

Supply Charges

____ A4216 Sterile water/saline/dextrose diluent/flush, 10 ml	____ A6213 Foam dressing, pad size 4x4 - 7x7, with adhesive
____ A4217 Sterile water/saline, 500 ml	____ A6214 Foam dressing, pad size 7x7 or larger, with adhesive
____ A4248 Chloreprep, 1 ml	____ A6219 Gauze, sterile, pad size 4x4 or less, with adhesive
____ A4332 Lubricant, individual sterile packet, each	____ A6220 Gauze, sterile, pad size 4x4 - 7x7, with adhesive
____ A4605 Tracheal suction catheter, closed system, each	____ A6223 Gauze, impregnated, vaseline, 4x4 - 7x7, no adhesive
____ A4624 Tracheal suction catheter, other than closed, each	____ A6231 Hydrogel, gauze, sterile, pad size 4x4 or less
____ A6196 Aquacell (alginate) dressing, 2x2	____ A6232 Hydrogel, gauze,, sterile, pad size 4x4 - 7x7
____ A6197 Enluxtra (alginate) dressing 5x5	____ A6233 Hydrogel, gauze, sterile, pad size more than 7x7
____ A6198 Alginate/fiber gelling dressing, larger than 7x7	____ A6234 Hydrocolloid (curafoam) dressing, 4x4 or less, no adhesive
____ A6199 Alginate/ fiber gelling wound filler, per 6 inches	____ A6235 Hydrocolloid dressing, pad size 4x4 - 7x7, no adhesive
____ A6203 Telfa (Composite) dressing, pad size 4x4 or less	____ A6236 Hydrocolloid dressing, pad size 7x7 or more, no adhesive
____ A6206 Mepitel One (Contact layer), 4x4 or smaller	____ A6250 Any skin sealants, protectants, moisturizers, ointments
____ A6207 Contact layer, sterile, between 4x4 and 7x7	____ A6255 Specialty absorptive dressing, pad size 4x4 - 7x7, with adhesive
____ A6208 Contact layer, sterile, larger than 7x7	____ A6257 Transparent film (tegaderm), sterile, 4x4 or less
____ A6209 Hydrocell (foam) dressing, 4x4 or less, no adhesive	____ A6258 Transparent film (tegaderm), sterile, 4x4 - 7x7
____ A6212 Mepilex (foam dressing), 4x4 or less, with adhesive	____ A6446 Kerlix (any size), each yard

Kanban Card Example

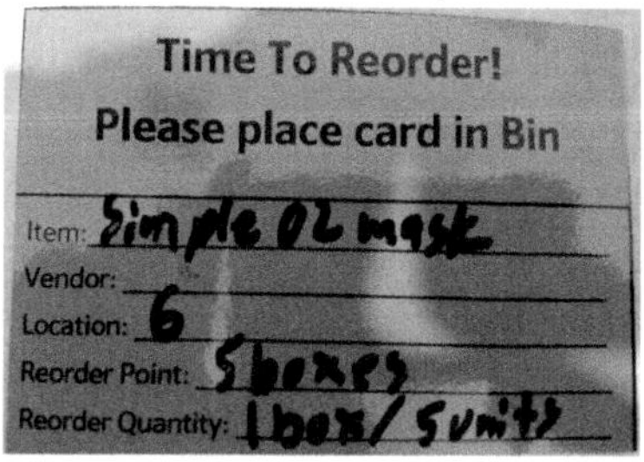

Shelf Labeling System

Kanban Reorder bin and Inpatient Charge Sheets

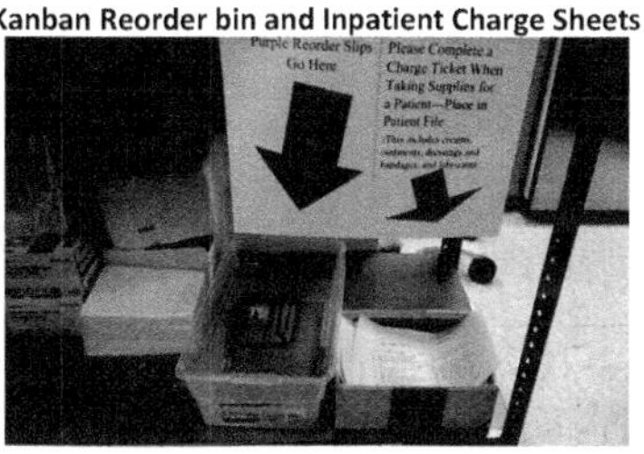

Large item (e.g. Catheter) Kanbans

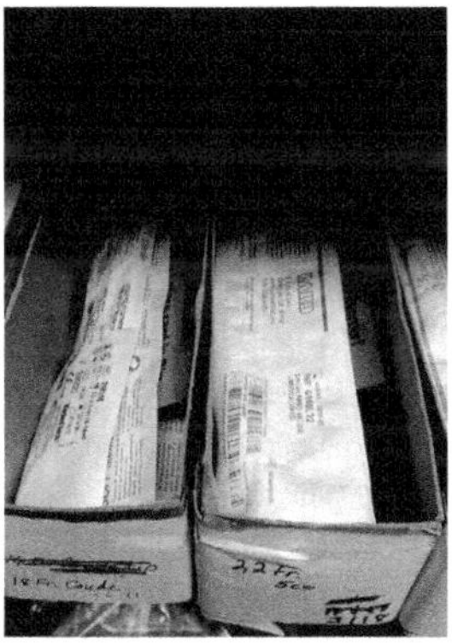

Small item (e.g. stopcock) Kanbans

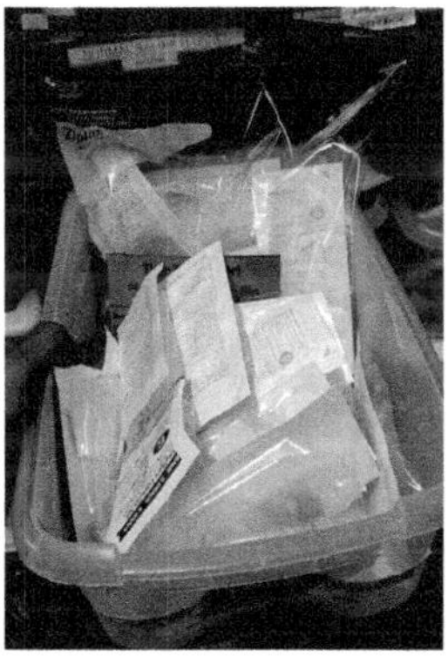

Updated Emergency Room Charge Sheet (Requested during Implementation)

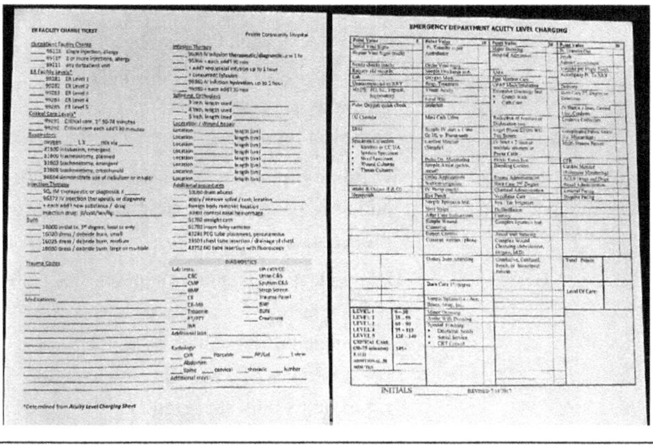

- Standardized process created, approved, and adopted by nurses.
- Charge Sheet created. Unable to estimate how much time nursing will spend managing inventory with the new inventory system.
    - Initially, probably more time-consuming due to the learning curve. However, it should be quicker for floor nurses to place orders.
- Approximately 8,100 individual items on hand (10% reduction), maximum, after the inventory system stabilizes.
    - Should be physically counted in October, January, and June to verify (see Follow-up Actions).

Follow-Up Actions

Action	Responsible Party	To Complete Action	Due
Evaluate the amount of storage space	Beryl	Visual check of shelves that look crowded or empty; inventory of the entire central supply.	October 20XX, January 20XX, June 20XX
Evaluate the amount of billable supplies being charged	Debbie	Compare $ amount of billable supplies purchased vs. estimated and actual $ of charged out supplies as detailed in the Financial Calculation document.	November 20XX, April 20XX, June 20XX
Spreadsheet maintenance	Beryl	Periodically add or delete items, adjust demand estimates, lead time, safety stock, and box quantity.	Weekly
Manage Kanban system	Beryl	Collect cards, place orders, replace cards, and make new cards as needed.	3 times/week
Evaluate facility charges and markups	Debbie	After the amount of billable supplies being charged has been evaluated, consider increasing facility fees to account for shortages or decrease supply markup if charged supplies far exceed the value of purchased supplies.	September 20XX
Labeling and organization of supplies	Beryl/Peter	After on-hand quantities stabilize, move shelves and supplies on shelves as needed to make best use of space. Label individual supply locations to preserve the organization scheme.	1/1/20XX–7/31/20XX
Review charge sheets	Chris/Beryl	Set times to review charge sheets to add or remove billable supplies/procedures.	Bi-monthly at trauma meetings
"Pink Sticker" system	Beryl	Date pink stickers, place them on items you suspect are not used/are not needed. If an item is used, the sticker comes off. Periodically get rid of items pink stickered.	9/1/20XX
Educate staff on new ER sheets	Beryl/Sherry	Orient staff on new ER charge sheets, specifically critical care and trauma sections.	7/31/20XX
Implement new ER sheets	Beryl/Sherry	Remove all old ER acuity and charge sheets from the ER; replace with new ER acuity and charge sheets.	7/31/20XX
Look into inventory management software	Debbie/Beryl	Look into different types of programs (e.g., Wasp, Peachtree, Fishbowl, etc.) to determine if cost-effective for this facility. Will require facility to keep a running tally of inventory.	December 20XX

Case Study 4.2: Inventory Organization, Storage, and Management System for Operating Room (OR)	Date: XX/XX/20XX
Sponsor: OR Department	Author: Kirk

Background

- The hospital has the mission to provide quality healthcare that serves our communities in a private, safe, and caring environment.
- OR nurses spend too much time looking for items that seem disorganized and located in multiple places.

Current Condition

Layout of Supplies by Location in OR Area

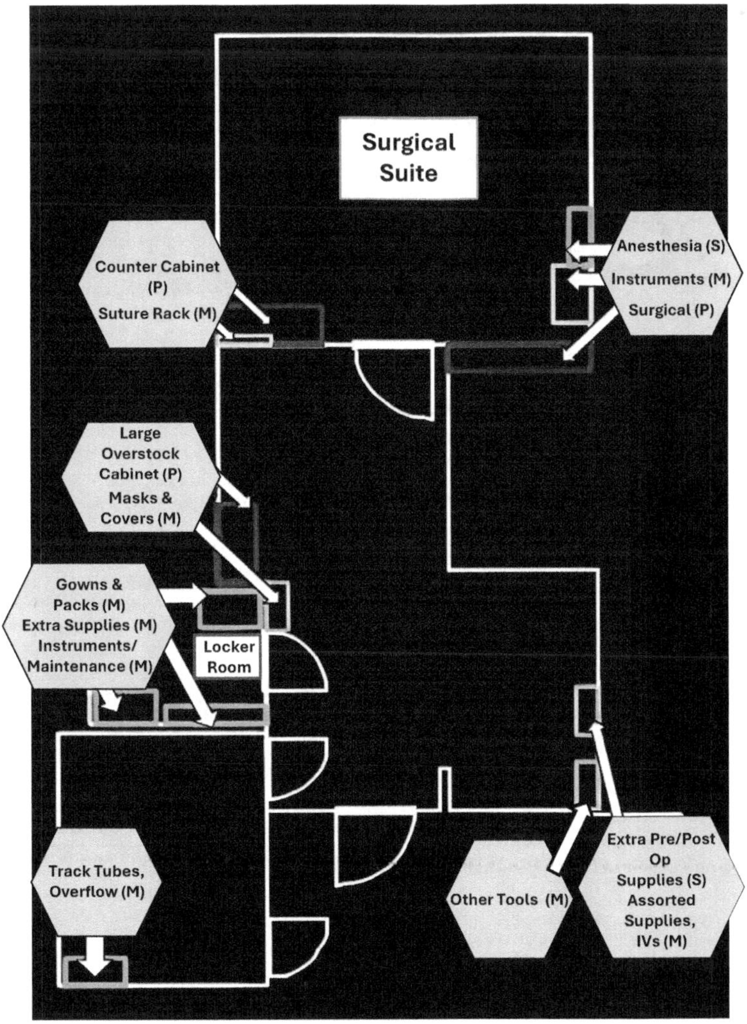

Example of Current State Cabinet/Shelf Layout

"C" cabinet "A" Shelf

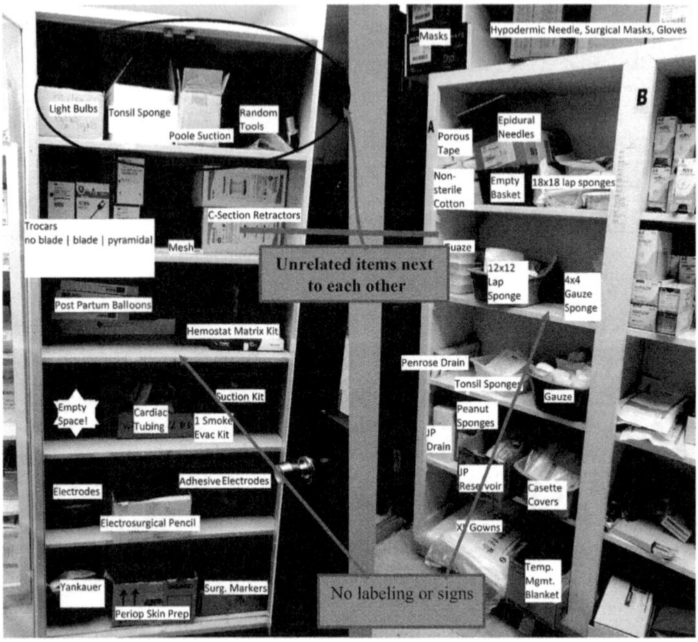

Search Time Experiment: With help from CON, a realistic pick list for an operation was developed. Three relatively new nurses were asked to find the items in a random order, one item at a time. Times were collected for each item (in seconds below).

Item	Nurse 1	Nurse 2	Nurse 3	Average
Water Trap	105	Not found	Not found	135
Suction Yankauer	9	11	10	10
Lap Sponges	50	17	21	29.3
Cautery Pencil	23	30	15	22.7
Bair Hugger	140	16	7	54.3
5.5 Oral RAE Tube, Uncuffed	110	13	47	56.7
20 Gauge Hypodermic Needle, 1.5 in.	45	24	16	28
Maternity Pad	71	24	13	36
Total	**553**	**285**	**279**	**378.3**

Water trap not found twice! Located in the locker room cabinet.

SMART Goal:	• Reduce nurse search time in OR by 20%

Root Cause Analysis:

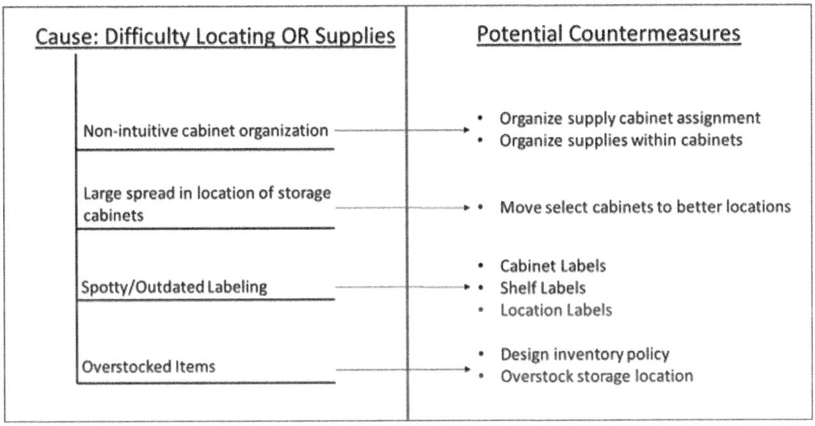

Proposed Counter Measures

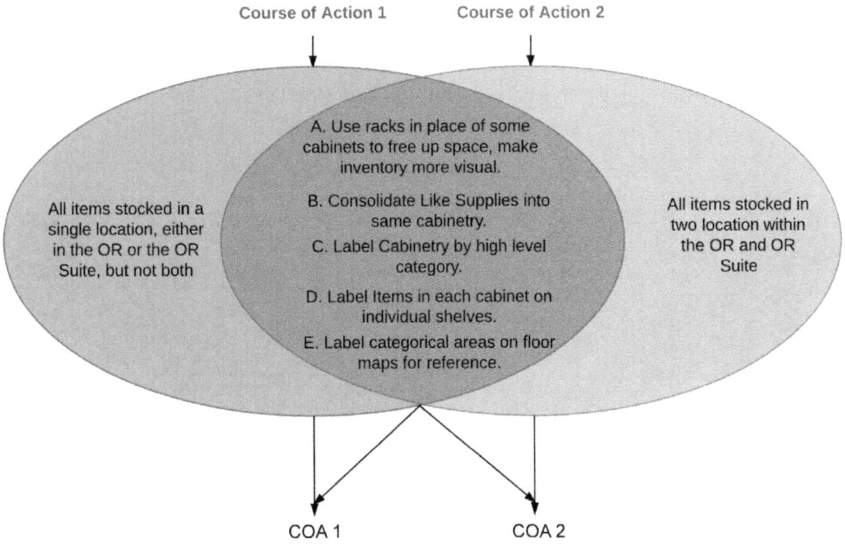

	COA 1	COA 2
Pros	• Simpler to stock. • Easier to find items in a single location. • Already have the space capacity in the current state. • Requires minimal layout changes to OR suite.	• Less total items in the OR Suite itself. • Less trips outside for nurses. • Like a high-level 2 bin Kanban system.
Cons	• May increase the number of trips outside the OR. • Requires the same number of bins as there are item singular item types.	• More work to stock, having every item in two places. • Requires additional cabinetry in the OR suite. • Affects a broader range of stakeholders than previously identified. • Requires two times the number of plastic bins when they are ordered.

Course of Action 1 was chosen by stakeholders

Proposed Labeling System for COA 1

- Item labels posted below all shelved items.
- Signs detailing contents for each rack and cabinet, color coded by high-level item grouping.
- Centrally posted map of high-level item grouping, represented by stars on the above floor plan.

Proposed Layout of OR post COA 1 Implementation

(Relocated cabinets circled in light grey, racks replacing cabinets circled in dark grey, map locations marked with stars)

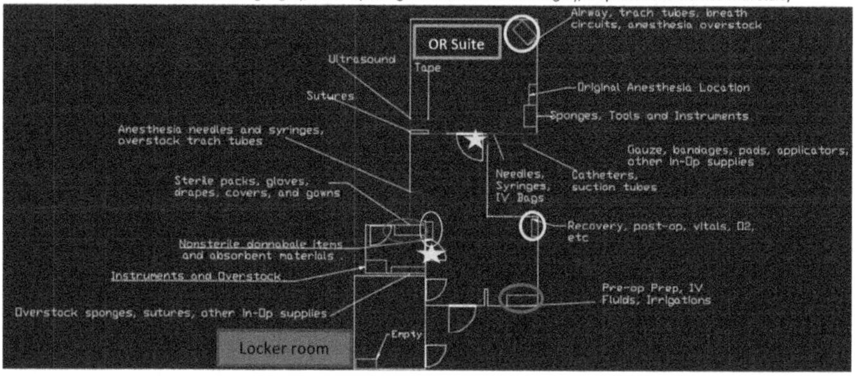

Implementation Plan

Action	Responsible	Deadline	Expected Outcome	Actual Outcome
Relocate racks/shelves within the OR	Ernesto/ Francine/ Facilities	Tuesday–Noon (week 2)	Racks inserted and cabinets removed.	Racks relocated to reflect the target state. Shelves delayed until 6/14 but were relocated.
Remove excess cabinets from OR	Ernesto/ Francine/ Gordon/ Facilities	Tuesday–Noon (week 2)	Workspace cleared up.	Excess cabinets removed.
Relocate OR storage items	Ernesto/ Francine/ Helene	Tuesday–Noon (week 2)	Items placed in new locations.	Items outside the OR suite relocated on schedule.
Relocate anesthesia items	Ernesto/ Francine/ Helene/Isaac	Tuesday–p.m. (week 2)	Anesthesia items placed in new locations.	Delayed until 6/14; anesthesia items placed in new locations.
Relocate items in the OR suite	Ernesto/ Francine/ Helene/Joyce	Wednesday–p.m. (week 2)	Determine obsolescent items, removing them. Rearrange and consolidate items.	Like items grouped and consolidated in the OR suite.
Types labels	Ernesto/ Francine	Wednesday–Noon (week 2)	Labels developed, printed, laminated.	Labels developed and printed by 2 p.m.
Apply labels	Ernesto/ Francine	Wednesday–p.m. (week 2)	Labels posted in new locations.	All cabinets labeled by item and by category.
Rearrange supplies as requested by stakeholders	Francine/ Ernesto	Wednesday–PM (week 2)	Have intuitive organizational scheme.	No additional requests after labeling implemented.
Reorient OR room	Helene	Thursday–a.m. (week 2)	Find new positions for the lamp and cautery generators and fridge, which were displaced by a cabinet.	Lamp, cautery generators, and fridge in suitable locations.

Effects Confirmation

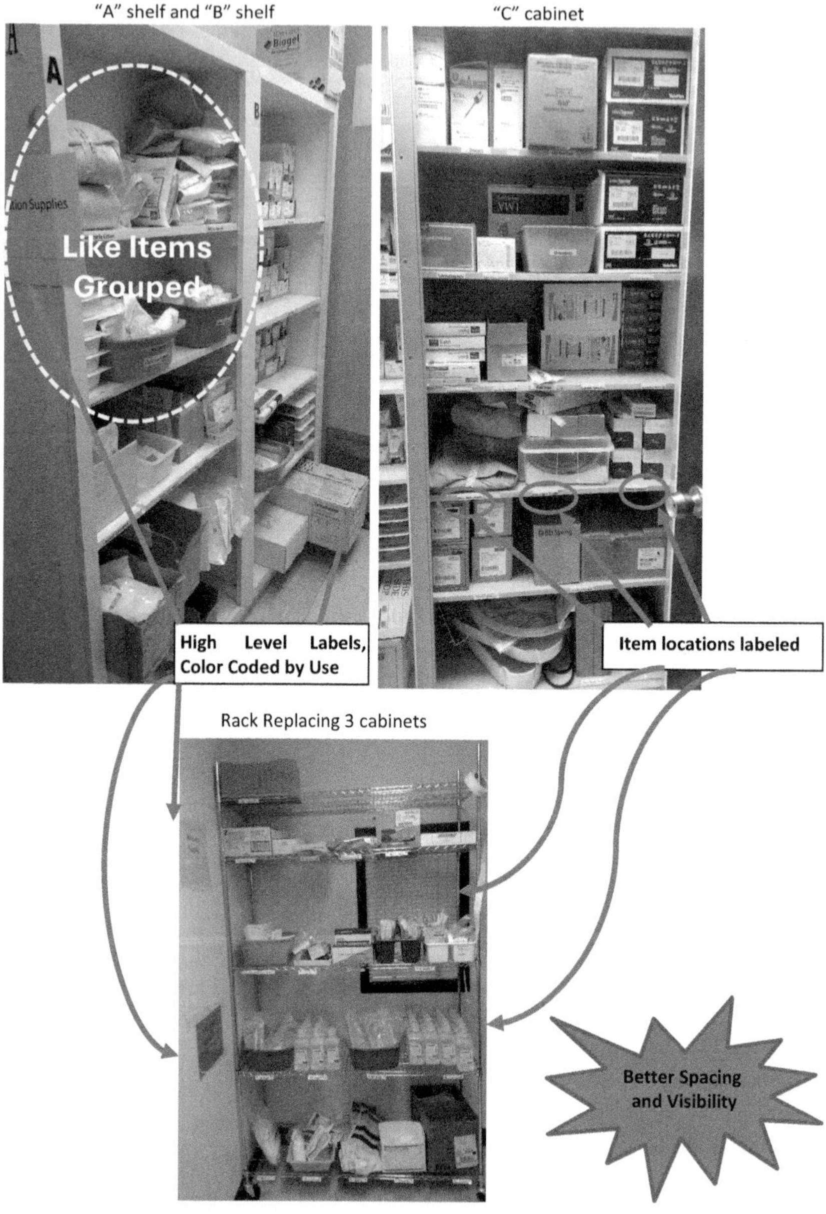

Item Name	Nurse 1 Search Time	Nurse 2 Search Time	Nurse 3 Search Time	Average Search Time	Notes
Water Check	150.00	150.00	111.43	137.14	*
Yankauer Suction	5.80	6.39	12.80	8.33	
Lap Sponge	3.84	13.42	6.05	7.77	
Electrosurgical Pencil	17.94	15.30	8.22	13.82	
Bair Hugger	31.00	14.65	150.00	65.22	***
Oral RAE tube, 5.5, Uncuffed	20.46	9.00	10.00	13.15	
20 G Needle, 1.5 in.	11.80	13.78	7.33	10.97	
Maternity Pad	90.00	50.38	53.83	64.74	**
Average	330.84	272.92	359.66	321.14	

* Feedback from both nurses 1 and 2 indicated that the sign wasn't helpful for locating them due to the area on the cabinet it occupied. An arrow was drawn to point to the location. If this doesn't help, consider relocating the water check to the 4′ Anesthesia Cabinet.

** Maternity Pads were found to be in multiple locations, consolidation occurred, and they are now in a single location and labeled as such.

*** Bair hugger time shows a negative percent improvement, but we received input for its location, counter to our knowledge. Removing this data point assuming the change had moved it to a more appropriate location, we see that this could improve to nearly 20% reduction in search time.

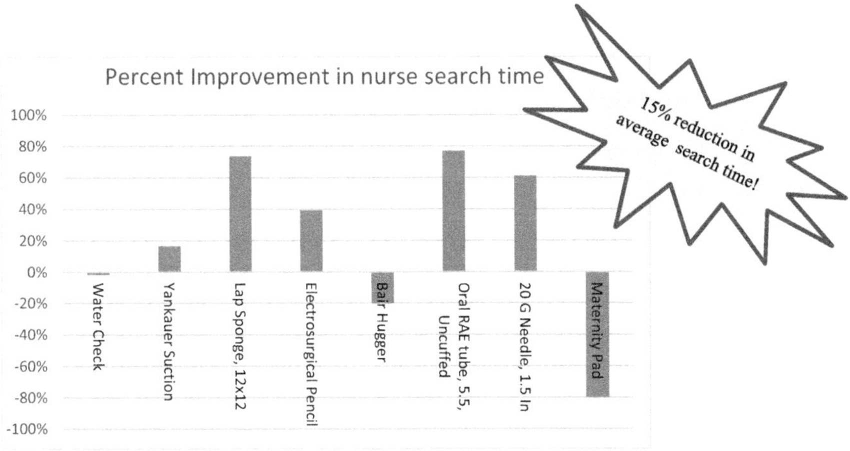

Follow-Up Actions

Action	Responsible Party	To Complete Action	Due
Weekly upkeep of system.	Gordon	Assign caretaker(s).	Ongoing
Expired cabinet.	Julio/Helene	Mount cabinet above counter, tape off sections for each department.	6/15/20XX
Repost sharps bin.	Work order by Helene	Decide on location and attach to wall.	6/15/20XX
Order plastic bins.	Helene	Count the number of bins needed of each size and place order.	7/07/20XX
Replace cardboard with plastic bins.	Work order by Helene	Pull all cardboard boxes out of OR.	7/21/20XX
Gauge effectiveness of an expired cabinet.	Helene	Develop a concise report/statement signifying how useful the cabinet is, and if it's a positive change as compared to the current set up.	8/15/20XX

Case Study 4.3: Improving Efficiency and Staff Access of Medical Supplies in the Emergency Department (ED)	Date: XX/XX/20XX
Sponsor: Emergency Department	Authors: Leslie and Nadine

Background

- Four emergency rooms.
- Each room has a specialized use.
  - Trauma Room, General ER Room, PEDS Room, Ortho Room (Splinting and Sutures).
- Each room stores supplies for the specific room use and common ER supplies.
- No place to store supplies outside of the treatment rooms.
- Two-bin system with set par levels for all inventory stored in the ED rooms.
- On-Hand Inventory is not based on actual usage.
- Staff may have to search multiple areas and rooms for supplies.
- Cluttered ED rooms.

Current Condition

A nurse located six items related to a peds patient with a lacerated head wound. Started in ED 01 (trauma room).

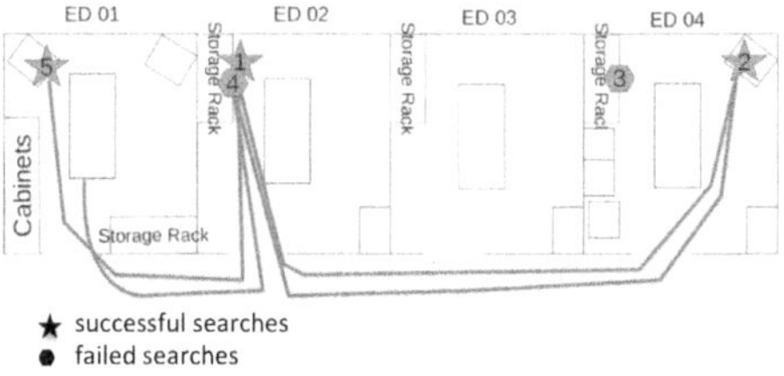

Nurse	Storage Units Searched	Distance Traveled	Time	Notes
Nurse	5	151 ft	2 min 7 sec	Patient in ED 03

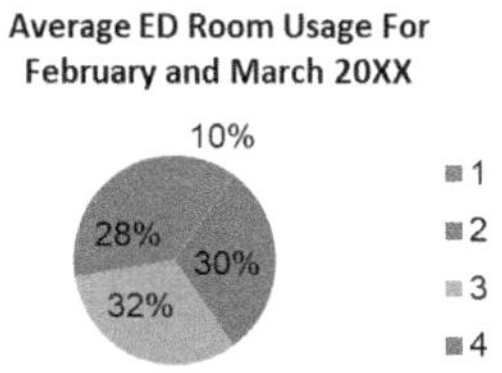

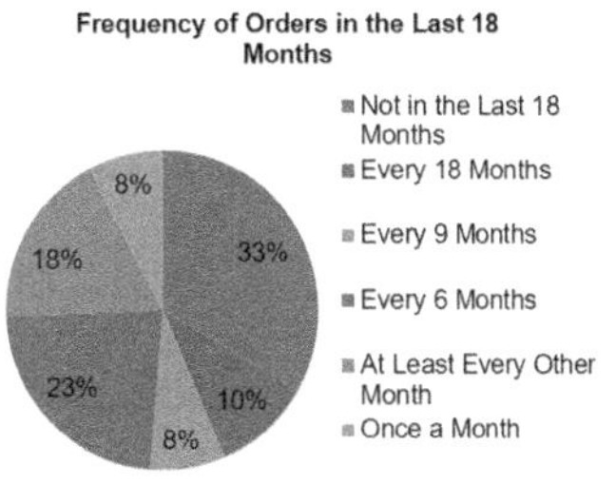

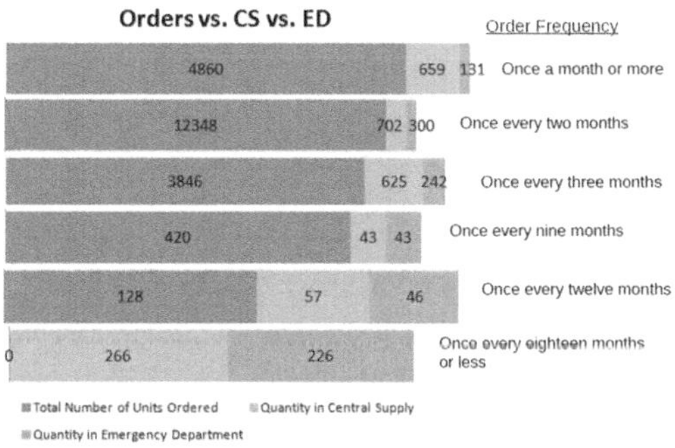

Item	CS Qty	ED Qty	2 Wk Usage	Use Category
Cathelon 20Gx1"	289	86	39	FAST MOVING
Oxygen Masks Adult	7	6	0	SLOW MOVING

SMART Goals:	
	• Reduce nurse search time to locate supply items in the ED by 50%. • Increase the amount of available space in each ED room for fast-moving supplies by 25%.

Root Cause Analysis

Goal 1:

- ➢Why is nurse search time so long in the Emergency department?
  - ➢Nurses have to travel to multiple rooms for supplies
    - ➢Why do nurses travel to multiple rooms for supplies?
      - ➢Ed rooms do not contain the same supplies
        - ➢Why do ED rooms not contain the same supplies?
          - ➢There is no room to keep all of the patient specific supplies in the different rooms
            - ➢Why is there no room?
              - ➢**The par levels are too high for some of the slower moving items resulting in area available for fast moving items**
- ➢Why is nurse search time so long in the Emergency department?
  - ➢Nurses have to search several areas in a room for supplies
    - ➢Why do nurses search several areas?
      - ➢The rooms have the same supplies placed in different spots
        - ➢Why do the rooms have the same supplies in different spots?
          - ➢Additional supplies were added as time progressed
            - ➢Why were additional supplies added?
              - ➢**The original location was not the most appropriate location**
              - ➢**New supplies were needed to be added**
              - ➢**Rack space was used for other items resulting in supplies being placed on any convenient surface**

Goal 2:

- ➢ Why is so much space taken up by inventory in the ED rooms?
  - ➢ Supplies can be found in multiple locations in each of the ED rooms.
    - ➢ Why are supplies located in multiple locations?
      - ➢ The item's initial location was conveniently placed.
      - ➢ The item could not be found and therefore was placed in an additional, easier-to-see location.
        - ➢ Why does this increase the amount of inventory in the rooms?
          - ➢ **A single item has 2 locations, so a higher quantity of that item is kept in the room than is necessary.**
- ➢ Why is so much space taken up by inventory in the ED rooms?
  - ➢ Par levels are the same for all inventory items.
  - ➢ There is no par level distinction between high and low usage items.
    - ➢ Why are all par levels the same?
      - ➢ **Par levels were based on item and bin size; larger items had smaller par levels.**
      - ➢ **Par levels were never actually tested in comparison to item usage data.**

Proposed Countermeasures

- Standardize shelving and cabinets in ED rooms.
- Remove excess inventory in ED.
- Remove durable medical equipment (DME) items from ED 4.
- Determine par levels for all ED inventory.
- Label all bins and shelving.

Implementation Plan

Action	Responsible	Deadline	Expected Outcome	Actual Outcome
Standardize shelving and cabinets in ED rooms.	Leslie, Nadine, Patty, Oscar, Rafael	Wednesday (week 2)	All ED rooms have the same items in the same location.	Expected outcome achieved.
Remove excess inventory in ED.	Leslie, Nadine	Wednesday (week 2)	Total ED inventory decreased.	Expected outcome achieved.

(*Continued*)

(Continued)

Action	Responsible	Deadline	Expected Outcome	Actual Outcome
Determine par levels for all ED inventory.	Leslie, Nadine, Patty, Oscar, Rafael	Wednesday (week 2)	Total ED inventory decreased; par levels based on usage data.	Expected outcome achieved.
Remove DME items from ED 4.	Nadine	Thursday (week 2)	Additional space opened up for shelf standardization.	Expected outcome achieved.
Label all bins and shelving.	Leslie	Thursday (week 2)	Supplies are easier to locate.	Expected outcome achieved.

Effects Confirmation

- Shelves and cabinets standardized.
- DME items relocated.
- Reduced total inventory.
- Reduced number of locations for inventory.
- Re-evaluated par levels.

A nurse located six items related to a peds patient with a lacerated head wound. Started in ED 01 (trauma room).

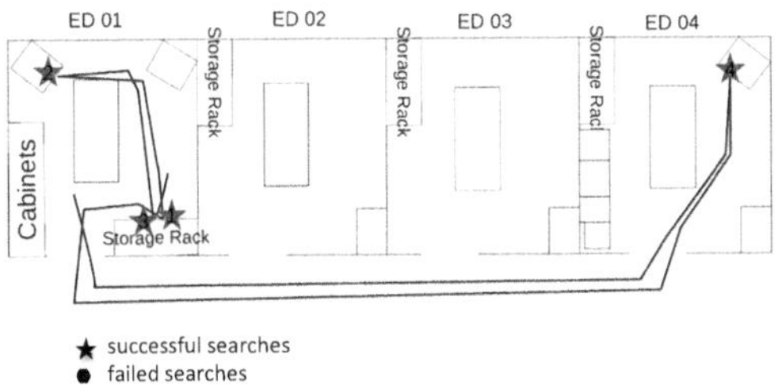

Nurse	Storage Units Searched	Distance Traveled	Time
Nurse 2	4	119 ft	1 min 3 sec

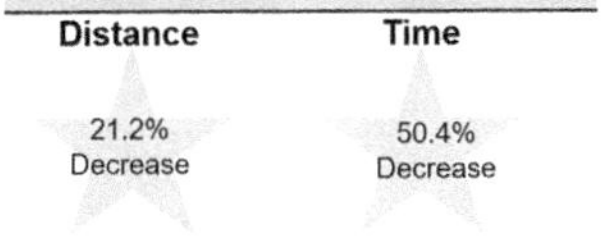

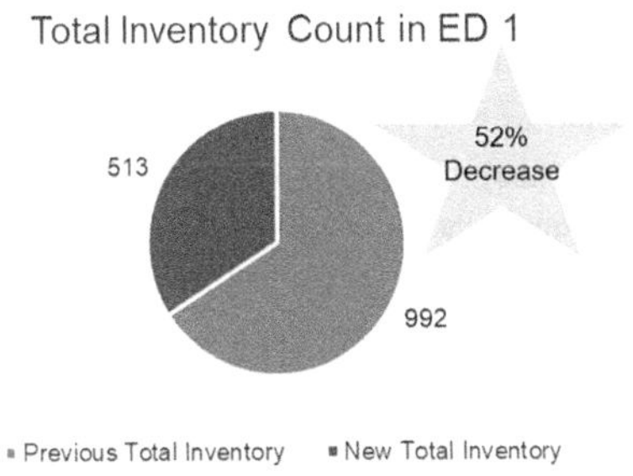

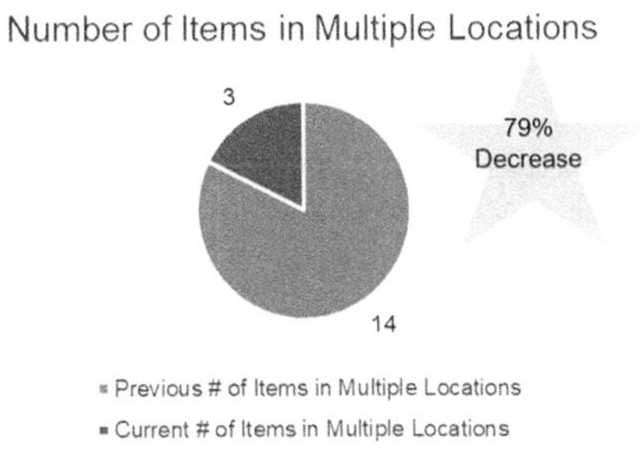

Follow-Up Actions

Action	Responsible Party	To Complete Action	Due
Place dividers in necessary bins.	Patty	• Bins that need dividers for each room. • Gloves (both bins). • Cath tap and wee bags. • Salem Sumps. • 20 mL and 30 mL syringes. • 60 mL syringes. • Tongue depressors and cotton tip applicators.	2 weeks.
Find a more appropriate cart for the ear, nose, throat, and eye cart in ED 4.	Patty, Oscar, Rafael	• A smaller cart that would take up less space.	2 weeks.
Train ED staff on new location of supplies.	Rafael	• Go over the supply list and location numbering system. • Physically look at ED inventory while looking over the location numbering system.	Next 3 weeks.
Match Nursing Station supply locations to standardized shelves in ED.	Patty, Oscar, nurses	• Use the same location numbering system.	3 weeks.
Reduce the amount of inventory in the Nursing Station.	Patty, Oscar	• Re-evaluate par levels based on usage.	Next 3 weeks.
Color coordinate the bins in ED 4.	Oscar	• Order more blue bins.	Next 3 weeks.
Clear the top of the carts.	Patty	• Get rid of any supplies on top of the carts. • Find a place in the rooms for the supplies OR store the items in central supply.	Next 3 weeks.
Re-evaluate the new location of the DME products.	Patty, Rafael	• Determine whether keeping the items inside the equipment closet is convenient or not.	Next 3 weeks.
Standardize the IV caddies in each room.	Oscar, Patty	• Purchase new caddies of the same size and design. • Place the same type and quantity of items in the same location in the caddies. • Add the caddy supplies and par levels to the Access database.	Next 3 weeks.

(*Continued*)

(*Continued*)

Action	Responsible Party	To Complete Action	Due
Evaluate the new system while completing inventory.	Patty, Oscar, Rafael	• Determine whether locations of supplies are convenient. • Determine whether par levels are adequate (are there enough of an item, are there too many of another item, etc.). • Compare reported out inventory total to the new, current inventory total (the inventory total at the beginning of July).	In 4 weeks.
Evaluate nurse flow in Med-Surg Department.	Nursing staff, Rafael, Patty	• Track distance and time while delivering medication and supplies to patients. • Determine whether there is a more appropriate centralized location for supplies in Med-Surg.	In 5 weeks.

Case Study 4.4: Improving Inventory Organization at Purchasing Department	**Date: XX/XX/20XX**
Sponsor: Purchasing Department	**Authors: Arlene and Cindy**

Background

- The purchasing department has just relocated to a new location.
- Purchasing department keeps supplies for the entire hospital. The only medication they are in charge of is over the counter drugs.
- The purchasing department replenishes the Medical Center every Monday, Wednesday, and Friday.

Current Condition

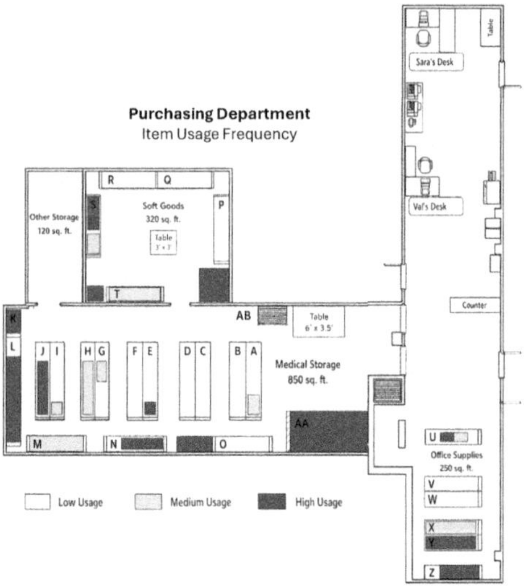

Shelving Unit	Category of Product
A	Urology Respiratory General Supply Suction-Stomach
B	Wound Care
C	Wound Care
D	Compression Stocking Surgery
E	Miscellaneous Endoscope
F	Casting
G	Tubing-Stomach Surgery
H	Hygiene Cups
I	Wound Care Urology Circulatory
J	IV Central Supply Hospital
K	OTC
L	IV Respiratory
M	Respiratory
N	Surgery Respiratory Stomach Urology Emergency Kit
O	Gloves
P	Immobilizers
Q	Immobilizers
R	Immobilizers
S	Custodial Supplies
T	Custodial Supplies
U	Miscellaneous /Office Supplies
V	Printing Supplies
W	Labels and Folders
X	Envelopes
Y	Paper
Z	Hospital Forms
AA	Adult Brief Liners & Accessories
AB	Candy

Item frequency current state.

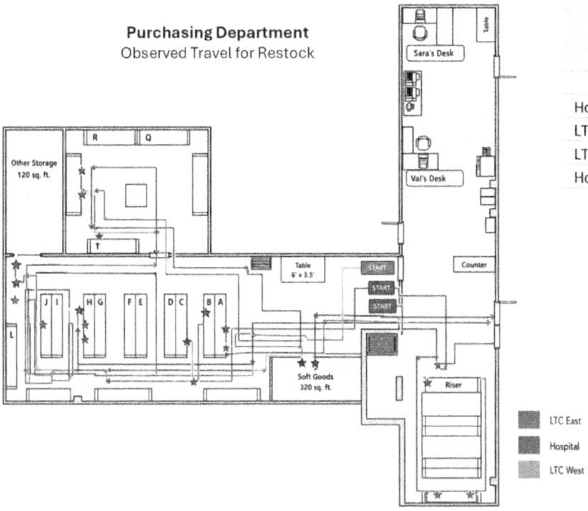

Observed pick distance of Purchasing Dept Employee (route showin Fig. 2)

	Travelled Distance (ft)
Housekeeping	148
LTC West Hall	136
LTC East Hall	427
Hospital	312

Search times for six randomly selected items by three users under the current inventory organization

User #1		9 min 21 s ec
User #2		7 min 1 sec
User #3		9 min 9 sec
	Average	**8 min 24 sec**

Observed travel for restock current state.

SMART Goals:	• **Reduce the Purchasing Department's travel distance by at least 50%.** • **Reduce average search time for any person who removes inventory from the department by more than 50%.**

Root Cause Analysis

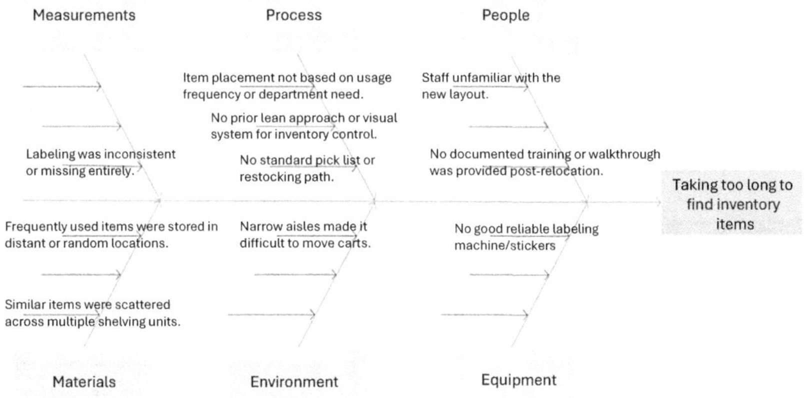

Why is it taking too long to find inventory items?

> Because items are not logically grouped or consistently labeled across shelving units.

Why are items not logically grouped or labeled?

> Because the inventory was not reorganized after the Purchasing Department relocated.

Why was the inventory not reorganized after the move?

> Because there was no clear inventory management system or layout plan prior to or during the transition.

Why was there no system or plan?

> Because the inventory practices were based on legacy ad hoc arrangements, and there was no prior standardization or process ownership.

Why was standardization lacking?

> Because of limited training, the absence of visual management tools, and insufficient attention to lean principles in materials management.

Proposed Countermeasures

1. Categorize all inventory by
   a. Department.
   b. Usage.
2. Move the frequently used items to the front.
3. Adjust shelving units to allow a cart to pass through the aisles.
4. Create a "grocery store" labeling system.
   a. Clear labels to mark the category of the aisle.
   b. Clear label to show what is in the aisle.
   c. Clear labels for items and sizes.
   d. Barriers for items.
5. Organize the soft goods room
   a. Body type.
   b. Create Aisles.
6. Create a map for the inventory to place at the front
   a. What aisle the category is.
   b. Where the aisle is located in the department.
7. Organize the pick list to match the new inventory layout.

Category Labeling of Aisle:

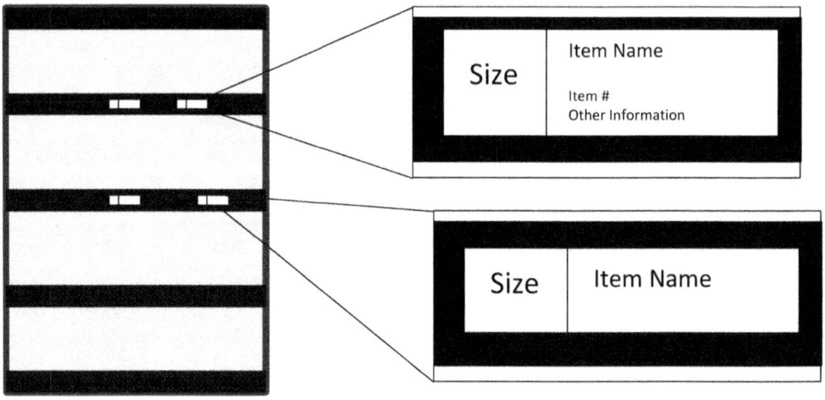

Possible Solutions for Labeling Items and Aisle

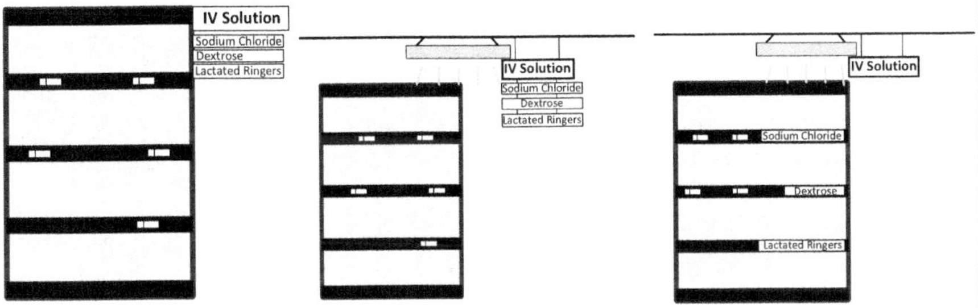

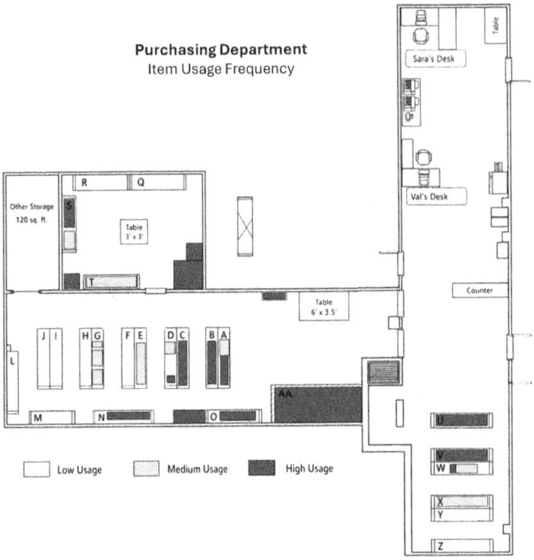

Shelving Unit	Category of Product
A	LTC Supplies
B	IV Supplies
C	IV Supplies
D	Syringes
E	Respiratory
F	Wound Care
G	Wound Care
H	Urology
I	Stomach
J	General / Central Supply
K	OTC
L	General / Central Supply
M	ER
N	Clinic
O	Gloves
P	Paper Products
Q	Immobilizers/Casting
R	Immobilizers / Casting
S	Custodial Supplies
T	Custodial Supplies
U	Forms
V	Paper
W	Misc Office Supplies
X	Envelopes
Y	Printing Supplies
Z	Labels and Folders
AA	Adult Brief Liners and Accessories
AB	Candy

Plan for item new location by frequency.

Implementation Plan

Action	Responsible	Deadline	Expected Outcome	Actual Outcome
Categorize all items	Arlene, Cindy, and Bret	Monday (week 2)	Items categorized by department.	Expected outcome achieved.
Plan category location	Arlene and Cindy	Monday (week 2)	Categories inserted on the layout map.	Expected outcome achieved.
Detailed storage implementation plan	Arlene and Cindy	Monday (week 2)	Item location input on the excel item database.	Expected outcome achieved.
Review detailed storage plan	Arlene, Cindy, Bret	Monday (week 2)	Plan reviewed and accepted by Bret.	Expected outcome achieved.
Relocate inventory to a new location	Arlene and Cindy	Monday–Tuesday (week 2)	Items moved into a new category location.	Expected outcome achieved.

(*Continued*)

(Continued)

Action	Responsible	Deadline	Expected Outcome	Actual Outcome
Label storage locations	Arlene and Cindy	Tuesday (week 2)	Most locations labeled.	Ran out of labeling supplies.
Pilot run and collect travel and search data	Purchasing	Wednesday–Thursday (week 2)	Pilot flow and search observed.	Expected outcome achieved.
Review new plan	Arlene, Cindy, Bret	Thursday (week 2)	Implementation reviewed and approved.	Expected outcome achieved.
Communicate results	Arlene, Cindy, Purchasing	Friday (week 2)	Implementation presented to the administration.	Expected outcome achieved.

Effects Confirmation

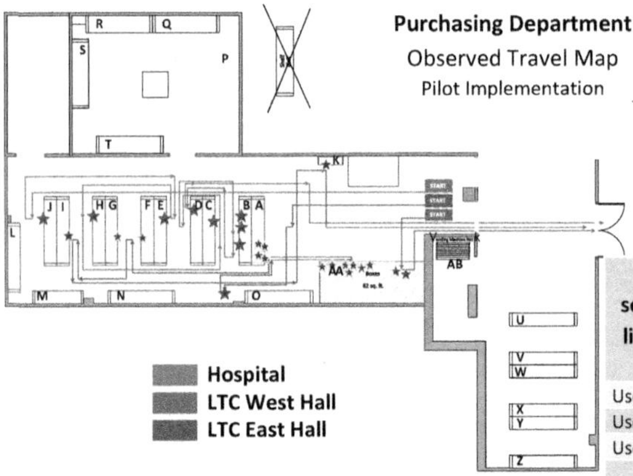

Search times for six randomly selected items by three users with little experience of new inventory organization	
User #1	1 min 38 sec
User #2	1 min 36 sec
User #3	2 min 2 sec
Average	1 min 45 sec

Observed pick distance of Purchasing Dept Employee after changes		
	Travelled Distance (ft)	Reduction (%)
LTC West Hall	32	76%
LTC East Hall	157	63%
Hospital	209	33%

57.33% reduction in average travel distance

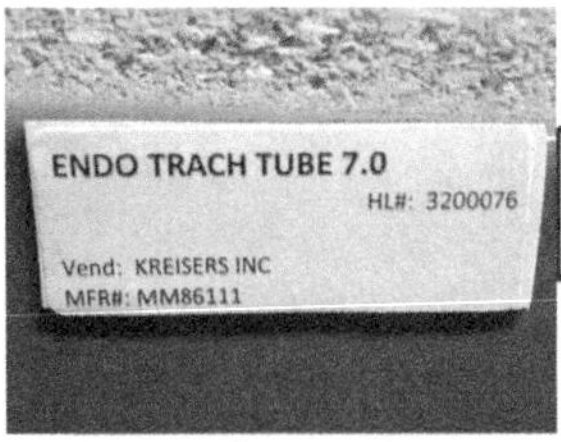

50% of storage locations were labeled.

Example of storage unit label

Follow-up Actions

Action	Responsible Party	To Complete Action	Due
Relocate office supply items to the designated area according to the plan.	Bret	Relocate office supply items to the designated area according to the plan.	Next 2 weeks.
Finish labeling storage locations with magnet labels.	Bret	Finish labeling storage locations with magnet labels.	Next 2 weeks.
Sequence the hospital restock list according to the stockroom layout.	Purchasing	Sequence the hospital restock list according to the stockroom layout.	In 3 weeks.
Organize department shelves throughout the hospital to match the restock list.	Purchasing, Bret	Organize department shelves throughout the hospital to match the restock list.	Ongoing.

Case Study 4.5: Personal Protective Equipment (PPE) Inventory Management	Date: XX/XX/20XX
Sponsor: Community Hospital	**Author: Emily**

Background

- XXXXX Community Hospital (XCH) is a critical access hospital that serves community members, local ranchers, and interstate travelers.
- During the COVID-19 Pandemic, they experienced an influx of medical supplies being sent to them by the government.
- These excess supplies caused a ripple effect of not knowing where to store them and not knowing exactly what supplies were on hand.
- This led to storing these supplies in a trailer that is supposed to be used for decontamination equipment.

Current Condition

Current State Value Stream Map (without timeline)

Note: The supplies in this inventory are only for emergency preparedness.

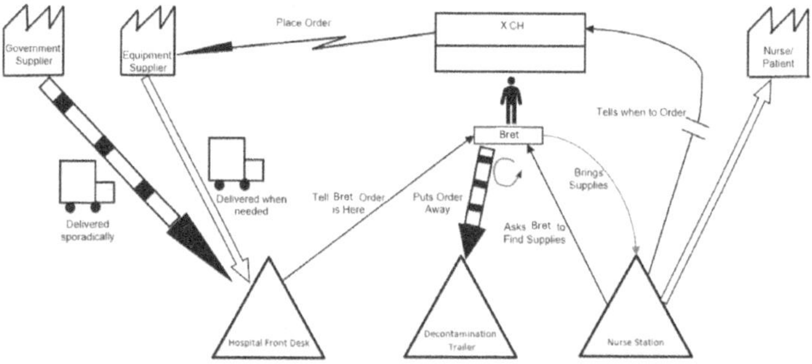

Trailer

Basement

Mask Inventory at XCH

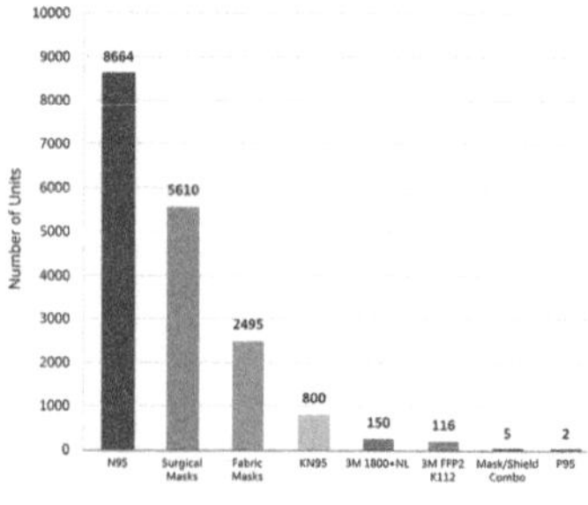

PPE Inventory at XCH

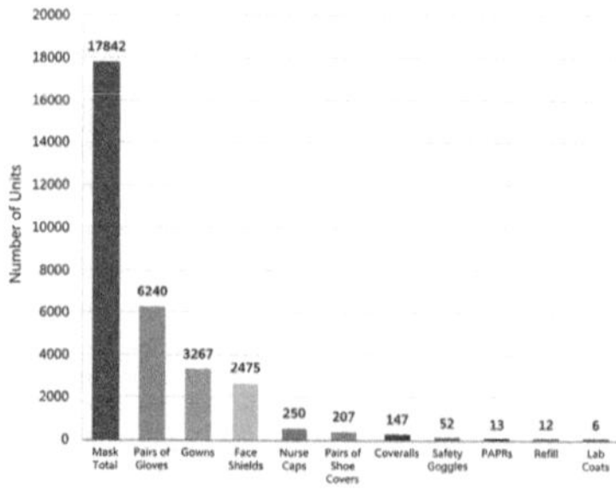

Major Problems:

- Inventory of PPE is undocumented.
- The trailer's original purpose is being compromised.
- The time it takes to find and deliver supplies, in the current condition, is at least 3 minutes.
- Excess amounts of PPE, with a total inventory of 30,523 items.

SMART Goals:	• **Create an inventory management documentation system that allows for PPEs to be found in less than one minute by anyone.** • **Reduce current on-hand emergency preparedness inventory by at least 25%.**

Root Cause Analysis

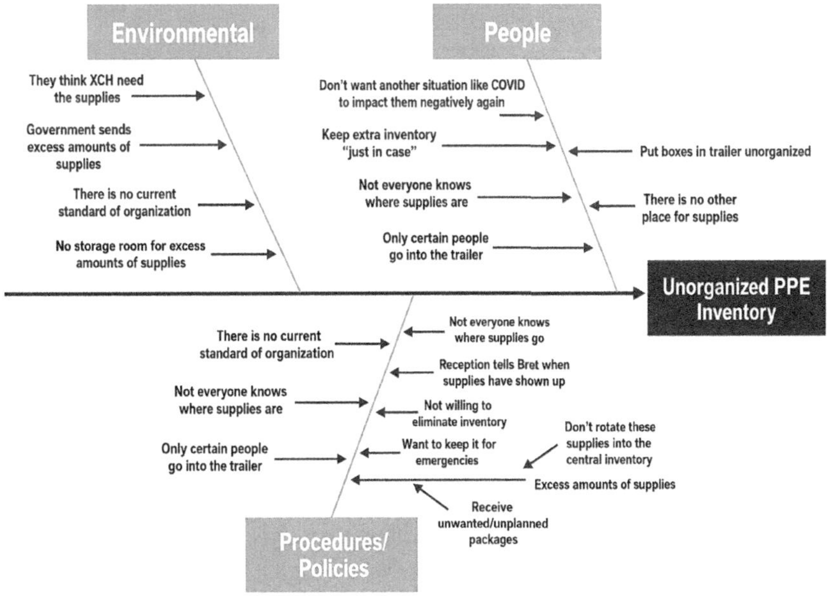

5 Whys:

Problem Statement 1: It takes too long to find PPE.

↳ **Why?** Only specific people can find supplies.

↳ **Why?** The trailer is unorganized.

↳ **Why?** XCH is receiving excess amounts of PPE.

↳ **Why?** The government is sending allocations.

↳ **Why?** No one has requested that they stop.

Problem Statement 2: It takes too long to find PPE.

↳ **Why?** Only specific people can find supplies.

↳ **Why?** The trailer is unorganized.

↳ **Why?** The trailer is too small to hold all supplies.

↳ **Why?** Supplies are put in the trailer without a formal organization plan.

↳ **Why?** They do not have a system of how to organize the supplies.

Problem Statement 3: Excess inventory of PPE.

↳ **Why?** There is no system to use the PPE before it expires.

↳ **Why?** They don't want to donate or discard any amount of the inventory.

↳ **Why?** They want to keep any and all amounts of PPE just in case of an emergency.

↳ **Why?** They don't want to run out of supplies in case another pandemic occurs.

↳ **Why?** They were unprepared when the COVID-19 Pandemic started.

Problem Statement 4: Excess inventory of PPE.

↳ **Why?** There is no system to use the PPE before it expires.

↳ **Why?** Have not considered their need or use for the PPE.

↳ **Why?** Think they need more PPE on hand now.

↳ **Why?** They fear they will run out of PPE if there is an increase in patients.

↳ **Why?** During the COVID-19 pandemic their use in PPE increased.

Proposed Countermeasures

- Build shelves and carry down all inventory.
- Donate any excess inventory.
- Talk with any person who uses the emergency equipment to create "frequently used together" guides.
- Organize inventory in a logical manner.
- Create a color-coded map of the inventory on the shelves so that supplies can be found quickly.
- Inform maintenance, emergency preparedness, nurses, clinic, and reception of the new changes.

TARGET CONDITION: Future State Value Stream Map (without timeline)

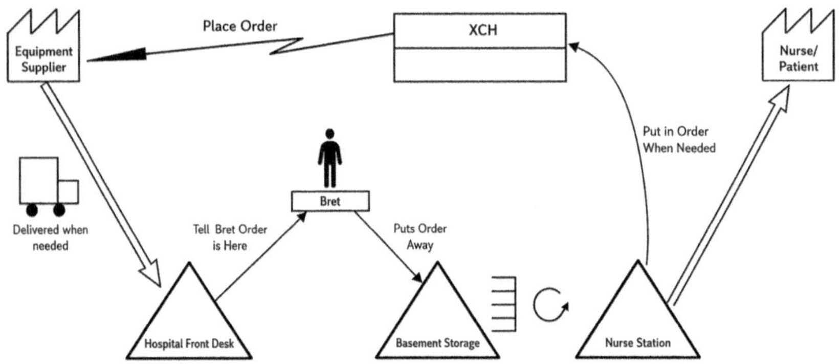

Implementation Plan

Counter Measures	Needed Actions	Responsible	Due	Expected Outcome
Incorporate current emergency inventory into everyday use with nurses and the clinic.	Talk with nurses and clinic about what supplies they may need right now.	Emily	Thursday (week 1)	• Reduce the amount of inventory.
	Give supplies needed to the appropriate department.	Emily	Friday (week 1)	
Donate PPE to others.	Research different locations in need of PPE.	Emily and Bret	Monday (week 2)	• Reduce the amount of PPE inventory.
Eliminate government supplier.	Have Bret get off the list of PPE allocations.	Emily and Bret	Monday (week 2)	• Reduce the amount of inventory coming to XCH.
Restock central supply.	Create emergency kits at nurse central supply with PPE.	Emily and Cindy	Monday (week 2)	• Nurses will be able to quickly and easily gown-up.
Move all PPE inventory out of the trailer and into the basement.	Move PPE.	Maintenance	Monday (week 2)	• Trailer can be used to store decontamination equipment again. • An organized layout of supplies.
	Count inventory.	Emily	Tuesday and Wednesday (week 2)	
	Install shelves.	Maintenance	Tuesday and Wednesday (week 2)	
Create an inventory management system to show stock numbers and reorder points.	Create a sign-out sheet for inventory, so Bret knows what supplies they have.	Emily	Monday (week 2)	• Reduce the amount of inventory. • Inventory numbers will be known at all times. • Allow staff to find what they are looking for.
	Talk to staff about PPE usage.	Emily	Monday and Tuesday (week 2)	
Create a map of items on the shelves and a list of commonly used items together for specific emergencies.	Get a list of commonly used emergency PPE items from Bret and Cindy.	Emily	Monday (week 2)	• Staff will be able to quickly find PPE supplies that they are looking for when an emergency arises.
	Once shelves are organized, create a map of the PPE on the shelves.	Emily	Wednesday and Thursday (week 2)	
	Label the PPE on the shelves with colors and shapes that correspond to the list and map.	Emily	Wednesday and Thursday (week 2)	

Effects Confirmation

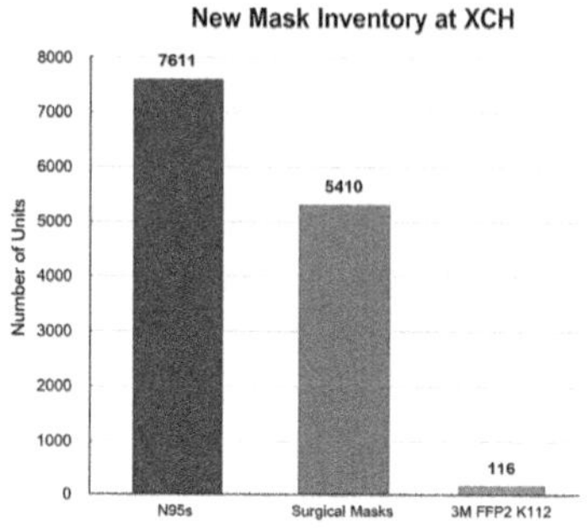

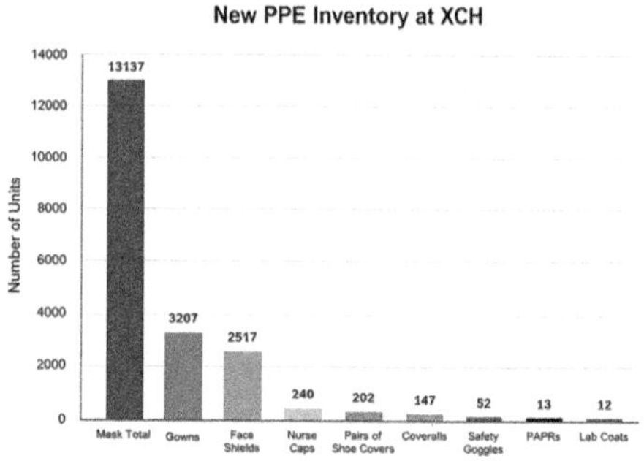

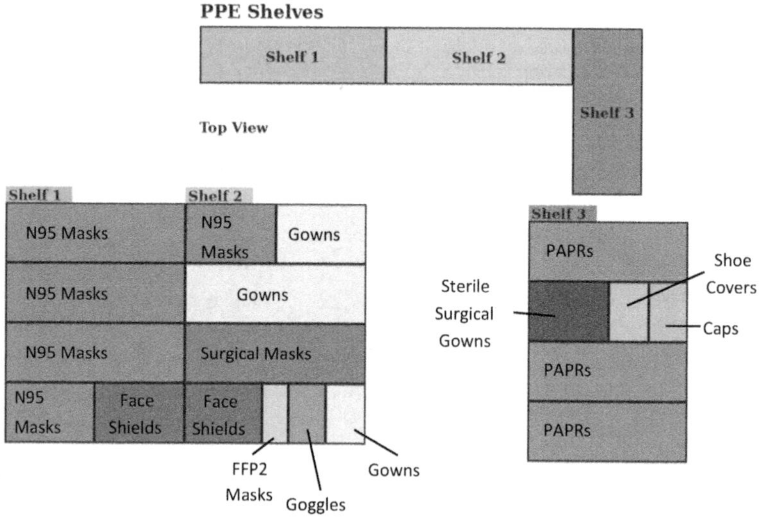

New Trailer

New Basement

Changes Made During Implementation:

- Nurse's supplies for emergency preparedness are now in their central inventory supply.
- Allowed nursing staff to stock expired N95s in their emergency preparedness stock.
- Trailer must be used to store PPE, because they declined to have PPE in a central location.

General Observations:

- Willing to donate supplies to other departments in the hospital to incorporate into everyday use.
- Not willing to donate any N95s.

Goals Met:

- **Nurses searching for inventory have been reduced to 10–20 seconds.**
- **Reduced total emergency preparedness inventory amounts by 36%.**
  - **Reduced mask inventory by 26%.**

There are still plenty of opportunities to reduce inventory!!!

Follow-Up Actions

Action	Responsible Party	To Complete Action	Due
Order new 3M 1860 N95s, because currently all the ones in stock are expired.	Bret and Cindy	Order new 3M 1860 N95s.	In 1 week.
Give Bret the sign-out sheet and map documents.	Bret	Map documents.	In 1 week.
Give Bret current PPE inventory numbers so he can update Juvare.	Bret	Update Juvare.	Next 3 weeks.

5 Clinic Workflow Case Studies

This set of RIE briefs illustrates continuous improvement accomplished through workflow redesign. Central to these efforts is thoroughly understanding the complexities of existing workflows through direct observation and data collection, carefully identifying activities that add value versus those which do not, and deeply investigating the underlying causes for the observed issues or problems. You will notice that in every case, the underlying causes point to specific elements within the workflow design (or lack of workflow design) and broader systems that interact with those workflows (e.g., scheduling systems or new employee training). The causal analyses never point the blame toward individuals, who are just doing their best to work within a flawed system.

The first case study deals with the workflows associated with scheduling appointments with a small radiology department. The current state workflow is quite complex resulting in delays for patients getting an appointment. By creating standard work around the "rules" for radiology appointments, nearly three-quarters of appointments could be made by the front desk without needing specialized expertise of the radiology technicians. This workflow improvement simultaneously improved patient satisfaction and reduced overall workload in the hospital.

Case 5.2 uses value stream mapping to identify value-adding and non-value-adding steps in the patient flow of an outpatient clinic. Through careful study of activity sequencing and coordination, the team was able to reduce patient in-process waiting time by 22% along with increased predictability of appointment length.

Case 5.3 addresses the challenge of notifying patients when lab results are received within a specified timeframe. When patients do not receive notification, their satisfaction is negatively impacted, and they frequently call the clinic which creates additional work for the staff and disrupts the work of nurses and providers. The countermeasures implemented addressed several root causes for delayed notification and improved the percentage of patients notified within 24 hours from 70% to 97%.

Case 5.4 presents a physical therapy department that was experiencing thousands of dollars of lost revenue per month due to lost charges because patient encounters did not get documented. After intense investigation of the workflow, several countermeasures were implemented that resulted in 100% of patient encounters being documented, i.e., elimination of lost revenue due to lack of a documented visit.

The final case study took on the challenge of increasing patient throughput by decreasing late arrivals and reducing patient waiting time. This case study illustrates the difficulty of addressing this type of problem in two weeks. While goals were met and the changes implemented moved metrics in a positive direction, the high

 DOI: 10.1201/9781003670698-5

opportunity countermeasures ended up listed in the Follow-up Actions section to be developed and implemented after the conclusion simply because they required more time than the two weeks allotted for the RIE.

As you review these cases, pay attention to how well the teams studied root causes, devised countermeasures to address them, and ensured the continuity of the implementation plans and effect confirmation sections.

Case Study 5.1: Reducing Radiology's Time to Schedule Appointments	Date: XX/XX/20XX
Sponsor: Outpatient Services	**Authors: Franklin and Gert**

Background

- Patients are referred to Radiology from both the WYZ Clinic and outside practices.
- Radiology consists of two technicians and has no dedicated receptionist.
  - Patient scheduling is done by the technicians.
  - Appointments are scheduled via the computer software, Health Land.
- When technicians are busy, patients are forced to leave a message, causing a delay in scheduling.
- The technician calls the patient back to schedule an appointment.

Current Condition

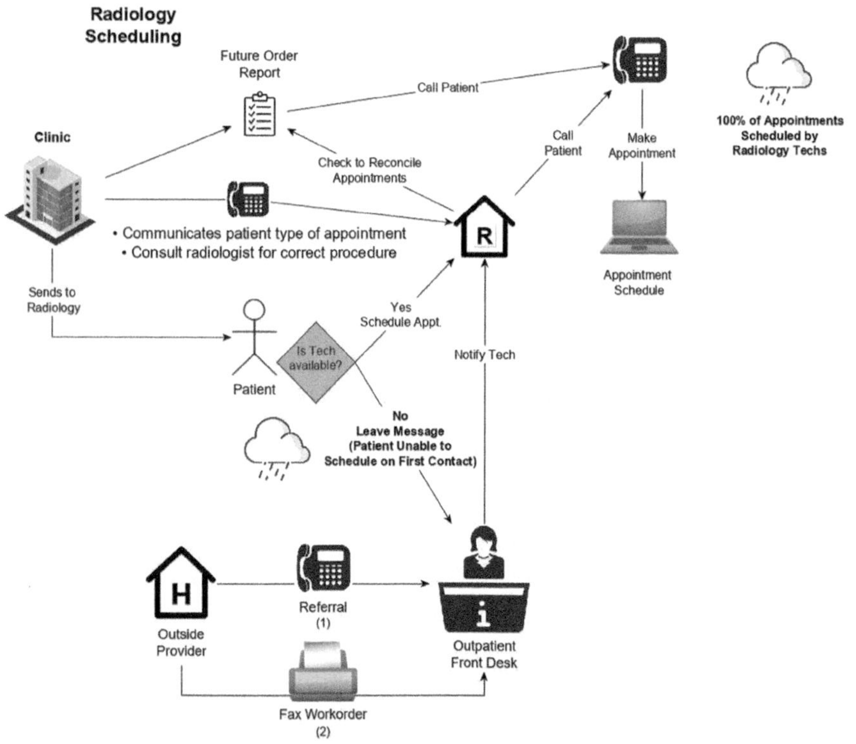

SMART Goals:	• Reduce the percentage of appointments scheduled by Radiology Techs to 30% or less. • 100% of patients are able to schedule an appointment on first contact (during business hours).

Root Cause Analysis

- **Radiology technicians are spending time on non-treatment** tasks.
  - **Why?** Radiology technicians are scheduling all imaging patients.
    - **Why?** Reception is unable to schedule for Radiology.
      - **Why?** Reception lacks knowledge of Radiology scheduling rules.
        - **Why?** No standardized scheduling rules have been established.
- **Patients unable to schedule an appointment on first contact.**
  - **Why?** Radiology technicians are busy treating patients.

Proposed Countermeasures

- Establish Scheduling Rules.
- Create Appointment Type Slip.
- Implement Rules in the Electronic Medical Record (EMR).

Implementation Plan:

Action	Responsible	Deadline	Expected Outcome	Actual Outcome
Phase I				
Establish scheduling rules	Franklin, Gert	Thursday (week 1)	Rules for Radiology Scheduling.	Rules created.
Create appointment type slip	Franklin, Gert	Thursday (week 1)	Communication method between the practitioner and reception.	Appointment slip created.
Phase II				
Implement rules in EMR	Harold	Tuesday (week 1)	Rules implemented.	Rules implemented.
Create five fake patient appointments for practice	Idalia	Thursday (week 2)	Train reception to use slip.	Expected outcome achieved.
EMR scheduling practice in clinic	Jose, Katia, Idalia, Clinic Staff	Friday (week 2)	Train reception in scheduling in EMR.	Expected outcome achieved.
Scheduling test	Jose, Katia, Idalia	Friday (week 2)	Reduce the percentage of appointments scheduled by Radiology Techs <30%.	70% of appointments were scheduled via front desk. Other 30% were future orders and special cases.

Effects Confirmation:

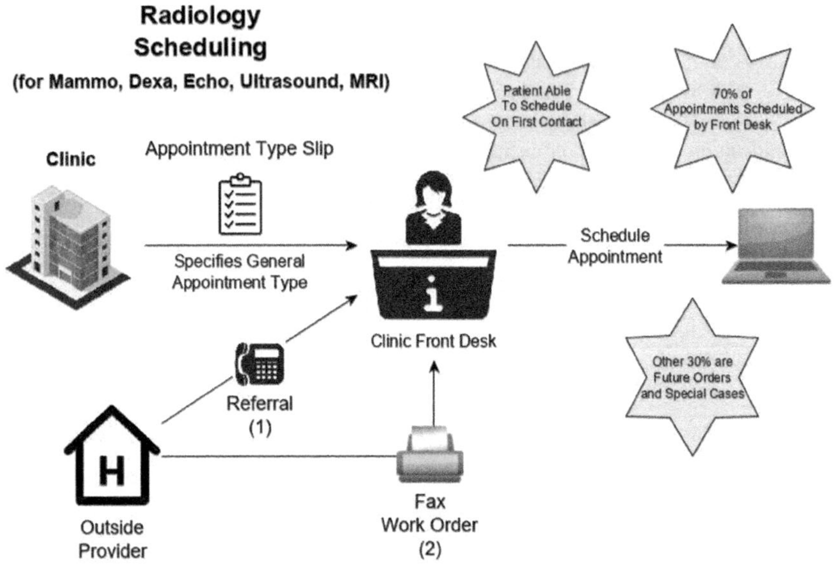

Follow-Up Actions

Action	Responsible Party	To Complete Action	Due
Measure percentage of appts. scheduled by front desk and Rad Techs.	Idalia		First of the month
Measure % of patients able to schedule on first contact.	Jose, Katia, and Idalia	Use the First Contact Tracker sheet.	First of the month

Case Study 5.2: Patient Flow Improvement	Date: XX/XX/20XX
Sponsor: Outpatient Services	**Author: Lee**

Background:

- A clinic sees an average of 11 patients per day.
- The clinic has six exam rooms, three on each side, and two providers:
  - Exam room 1 used most often.
  - Exam room 3 used least often.
- Appointment lengths are defaulted in the scheduling system:
  - Brief: 15 minutes
  - Regular: 20 minutes (physicals, suture removal, etc.)
  - Long: 30 minutes (procedures, family consultations, etc.)
  - Extended: 45 minutes (full yearly exams, chronic care, etc.)

Current Condition:

Issues:

- Patient visits are taking longer than expected.
- Visit times do not match the projected schedule; the clinic cannot see as many patients as is preferred.
- The discrepancies between scheduled visit time and actual visit time have not been identified.

Effects on Care:

- Cannot care for as many patients.
- Longer wait times frustrate patients.
- Can ultimately lower the standard of care being delivered to patients.

Patient Visit Simplified Value Stream Map

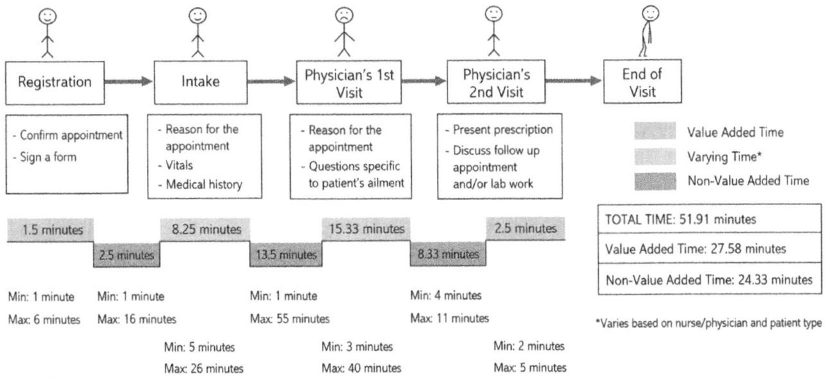

Patient Appointment Timeline for Observation Day

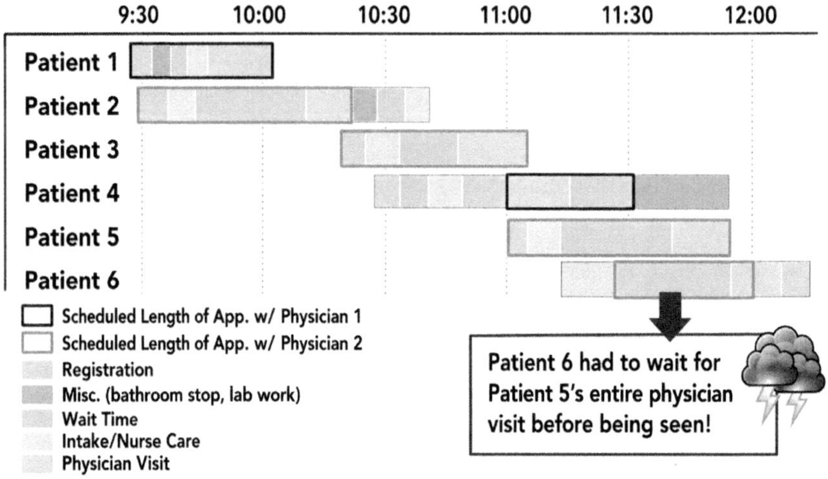

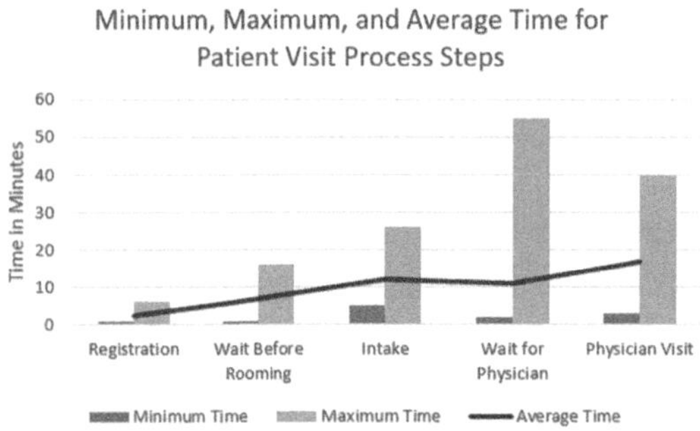

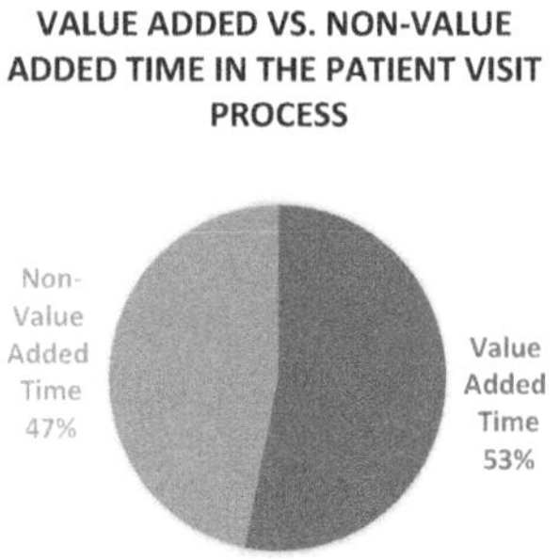

- ❖ Regular patient visits = 30 minutes.
- ❖ Average time for a regular patient visit over two days of observation = 48.42 minutes.
  - ❖ 18.42 minutes over the scheduled time

SMART Goals:	• Reduce patient wait time by 25%. • Reduce overall patient visit time by 25%. • Match expected length of patient visits with the actual length of patient visits.

Root Cause Analysis:

Why are patient appointments overlapping?

- The appointments are taking longer than expected.

Why are appointments taking so much longer than expected?

- Patients are spending a lot of time waiting to be seen by a physician.

Why do patients spend so much time waiting?

- The physician is busy seeing another patient.

Why are two patients ready to be seen by the same physician at the same time?

- Appointments run an average of 18 minutes over the scheduled time.

Why does this occur?

- The appointment is scheduled for the entire visit but does not account for wait time or intake time.
- The schedule is not separated into a nurse schedule and a physician schedule; therefore, the physician does not know when a nurse should be done with the intake and a patient should be ready to be seen.

- The physician does not know the patient is ready to be seen.

Why doesn't the physician know when the patient is ready to be seen?

- There is no separate nurse and physician schedule, so the physician only has a vague idea of when the patient should be ready to be seen.
- The schedule may be so far behind the physician does not even have a vague idea of which patient is being seen and therefore when that patient should be ready to be seen.
- There is no indicator present to show whether a patient is waiting for the intake or is waiting to be seen by the physician.

- The physician does not know there is a patient in one of the rooms.

Why wouldn't a physician know a patient is in a room waiting to be seen?

- The nurse may have forgotten to place the patient's chart on the outside of the door.
- There is no clear indicator, other than the chart, telling the physician a patient is present.

Proposed Countermeasures

- Place a "Patient Status Indicator" outside each exam room.
- Adjust the scheduling process so patients arrive 15 minutes before the scheduled physician visit.
- Separate the schedule between nurse and physician time.
- Adjust defaulted appointment lengths according to patient visit data.

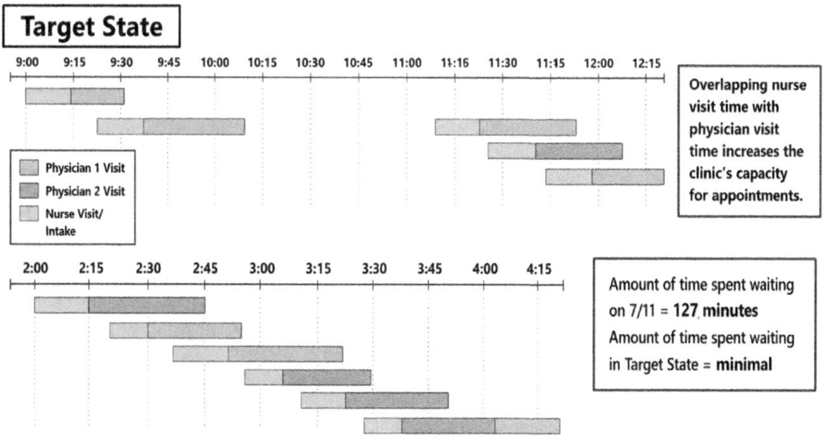

*Appointment data taken from one observation day

Physician Changes:

- Keeping the schedule open on a computer screen at all times.

Patient Status Indicator:

- Four different colored index cards stuck to the outside of the exam room door.
  - No index card = room is empty.
  - Blue index card = patient is being seen by the nurse.
  - Yellow index card = patient has been seen by the nurse and is ready to be seen by the physician.
  - Green index card = patient is being seen by the physician; knock, then wait for an answer before entering.
  - Red index card = patient is being seen by a physician and a procedure is in process; do not enter.
- **Continuous indicator of the patient status.**

Adjusting the Scheduling Process:

- Patient notified the appointment starts 15 minutes earlier than what the schedule indicates.
  - Example: The patient is told the appointment begins at 9:15; the physician schedule shows the appointment beginning at 9:30.
- Removes any responsibility from the patient to show up early for an appointment.
- Allots time for the nurse visit/intake.
- Gives physicians the full appointment time to see the patient.
- Allows appointments to begin to overlap nurse visit time and physician visit time.
- Schedule reflects physician time not including intake time.
- **Can increase the clinic's capacity to 32 appointments in 1 day.**
  - 30-minute appointments from 9 a.m. to 12 p.m., 20-minute appointments from 1 p.m. to 4 p.m.

Implementation Plan:

Action	Responsible	Deadline	Expected Outcome	Actual Outcome
Place a "Patient Status Indicator" outside each exam room.	Lee, Margot	Tuesday (week 2)	Clear indicator of whether a patient is present and ready to be seen by a physician or nurse.	Expected outcome achieved.
Adjust the scheduling process so patients arrive 15 minutes before the scheduled physician visit.	Lee, Margot, Nigel	Tuesday (week 2)	Patients arrive in time to be seen by the nurse. Schedule accommodates for nurse and physician time.	Expected outcome achieved.
Separate the schedule between nurse and physician time.	Margot, Nigel	Tuesday (week 2)	Physicians and nurses have a better understanding of the schedule as a whole. Physicians know when a patient has completed the intake and is ready to be seen.	Expected outcome achieved.
Adjust defaulted appointment lengths according to patient visit data.	Margot, Nigel	Wednesday (week 2)	More accurate prediction of the length of a patient's visit and the schedule as a whole.	Expected outcome achieved.

Effects Confirmation

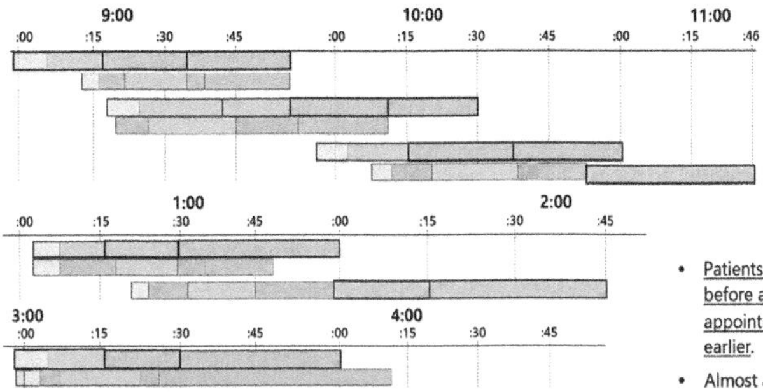

- Boxed appointments represent 'Expected' appointment lengths
- Each vertical line signifies 15 minutes

AVERAGE PATIENT VISIT TIMES	Previous System	New System	% Change
Time with Provider	16.84 minutes	21.72 minutes	28.9% increase
Patient Wait Time	17.7 minutes	13.8 minutes	22% decrease
Total Patient Visit Time	48.84 minutes	49.2 minutes	0.7% increase
Dif. Bet. Expected and Actual Visit Time	18.42 minutes	7.4 minutes	**59.8% decrease**

- Patients were called the day before and told their appointments started 15 minutes earlier.
- Almost all patients showed up 20-30 minutes early
- The average difference between the expected length of an appointment and actual length of an appointment was **7.4 minutes**

Follow-Up Actions

Action	Responsible Party	To Complete Action	Due
Train all staff involved on a new scheduling system.	Ophelia	Be sure nurses and physicians know the schedule reflects the times at which a patient should be ready to seen by a physician. Be sure all those who have the responsibility of scheduling appointments are aware of the new system.	Next 2 weeks.
Train all staff involved on a new note card system.	Ophelia, nursing staff, physicians	Review the meaning of each colored card. Make revisions to the colors if necessary (making the pink card red instead, etc.). Stress the importance of participation.	Next 2 weeks.
Develop a protocol for patients who have been waiting for an extended period of time.	All clinic staff	Keep the patient updated. If possible, make the longest wait be in the waiting room. If the patient is going to be waiting for 60+ minutes, provide the option to reschedule.	In 4 weeks.
Re-evaluate the new scheduling system.	Ophelia, Rina, all clinic staff	Continue collecting appointment timing cards. Analyze this data and use it to evaluate whether the system is actually working or not.	In 6 weeks.
Re-evaluate the note card system.	All clinic staff	Hold a meeting to discuss whether the cards are getting used or not (as well as why they are or aren't being used). Discuss ways to improve upon the note card system (more cards, fewer cards, etc.). If the card system is not working at all, discuss alternatives such as flags or lights outside the exam rooms.	In 6 weeks.

Case Study 5.3: Documented Ancillary Service Response	Date: XX/XX/20XX
Sponsor: Medical Center	**Authors: Rina and Philippe**

Background

- The Medical Center has experienced a lack of patient satisfaction because patients have not been contacted by their provider in an appropriate amount of time regarding their lab results from ancillary services (radiology, blood work, urine samples, etc.).
- If providers do not directly contact patients with results, then nurses may contact patients with the approval of a provider.
- If patients are not being contacted in a timely manner with their lab results, then they may not be aware of health conditions that could potentially harm them.
- If patients are not contacted by a provider regarding results within a business day and are not made aware of this by their provider, it results in excess calls to clinic reception that have to be routed back to nurses and providers.
  - Such calls disrupt appointment check in and scheduling and force staff to sift through voicemails before making calls back to patients.

Current Condition

- Data collection represents clinic visits during the first week of the Rapid Improvement Event.
- Data collection represents a normal week of visits from patients.
- Sample size is 143 patients who visited one of three primary care providers.
- Contact within 24 hours is represented by phone calls, in-person conversations during appointments, and letters sent during that same business day.

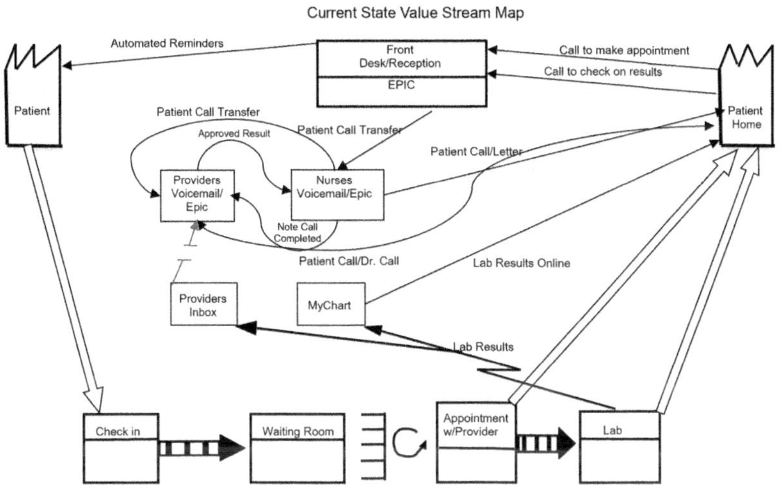

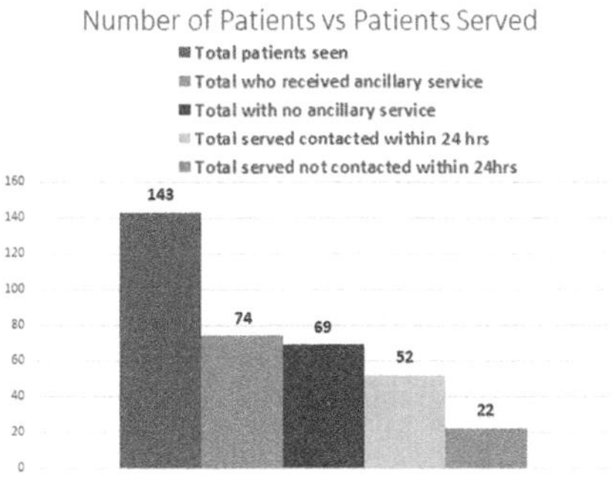

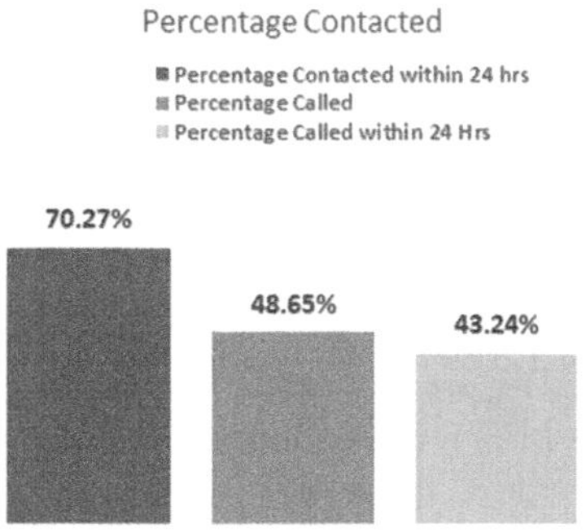

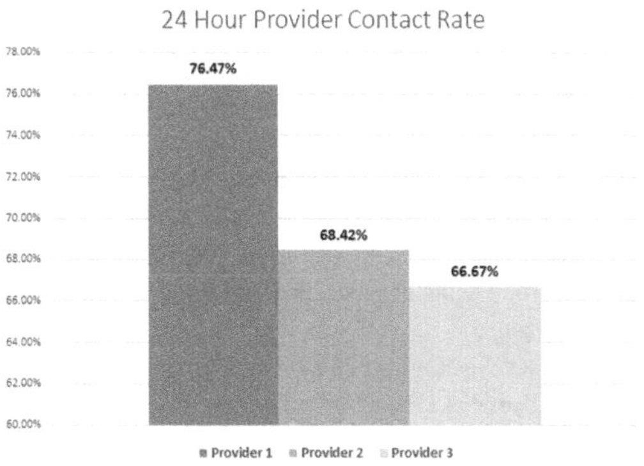

Major Problems

- Approximately 30% of patients are not contacted within 24 hours of their lab results becoming available.
- Approximately 27% of patients are not contacted with their lab results.

SMART Goals:

- Increase the percentage of patients who had ancillary services done, to be contacted within 24 hours of their lab results becoming available during a business week from 70.27% to at least 80%.
- Reduce the number of patients without documented contact from 27% to 17% or lower.

Root Cause Analysis

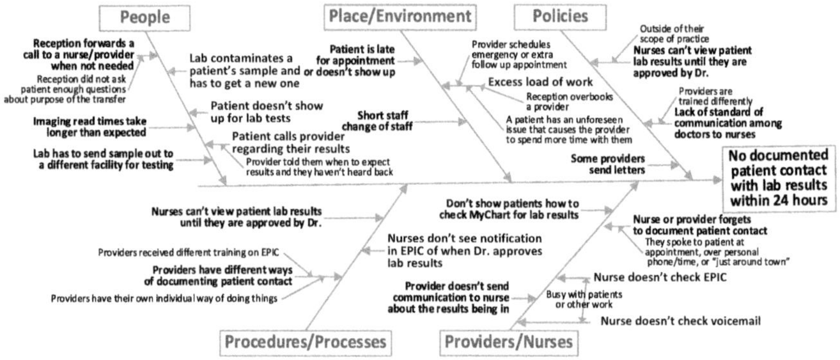

5 Whys:

Problem Statement 1: No documented point of contact with the patient.

↳ **Why?** The patient doesn't receive their results via letter, phone call, or during their clinic appointment.

↳ **Why?** Providers and nurses are overwhelmed by phone calls.

↳ **Why?** Front desk diverts unnecessary calls to nurses' and providers' inboxes.

↳ **Why?** Nurses and providers don't communicate with patients when they should expect their results.

↳ **Why?** Nurse and doctor communication with patients is not standardized.

↳ **Why?** Doctors and nurses have developed their own methods of practice.

Problem Statement 2: No documented point of contact with the patient.

↳ **Why?** The patient doesn't receive a letter, phone call, or in person to communicate results.

↳ **Why?** Providers and nurses are overwhelmed by phone calls.

↳ **Why?** Front desk receives an excess of phone calls from concerned patients.

↳ **Why?** Front desk diverts unnecessary calls to nurses and providers inboxes.

↳ **Why?** Reception fails to ask clarifying questions as to why a patient is calling.

↳ **Why?** They're not expected to do anything besides get the patient to where they want to go.

Problem Statement 3: Nurses don't see when providers approve a patient's results.

↳ **Why?** The provider did not make the nurse aware of a patient getting testing done.

↳ **Why?** The nurse is not aware of how to see when tests get ordered via Epic.

↳ **Why?** The nurse received different training on Epic.

↳ **Why?** Lack of standard when training new employees.

↳ **Why?** There's no dedicated training program.

Problem Statement 4: No documented point of contact with the patient.

↳ **Why?** The patient doesn't receive a letter, phone call, or in person to communicate results.

↳ **Why?** Nurses can't contact patient to explain results until doctor reviews results.

↳ **Why?** Outside nurses' scope of practice.

↳ **Why?** Laws governing healthcare that limit the scope of practice.

Problem Statement 5: Not returning patient calls.

↳ **Why?** Patient appointments are prioritized ahead of listening to all voicemails and returning patient calls.

↳ **Why?** There are too many voicemails and calls to return during their scheduled day.

↳ **Why?** The current time to listen to voicemails and respond is incompatible with their schedule.

↳ **Why?** Voicemails must be listened to in full and can last 1–5 minutes.

↳ **Why?** Providers must ensure they fully understand the patient's needs.

Proposed Countermeasures

- Reception Phone Call Flowchart.
- Lab to patient communication pamphlet.
- Epic software use training.
- Time-defined flow of reviewing results and calling patients between/during appointments depending on availability.

Target Condition:

Lab to Patient Communication Pamphlet

**Note similar sheets were made for 48 hours and 3-7 business days

Important things to note after your Lab today

❖ Your lab/radiology tests will be completed within 24 business hours

- Remember as soon as your lab is done it will appear in your MyChart results tab

❖ You can expect to hear back from your provider during the business week by phone or mail* within 24 hours after the lab has finished processing your info.

*Note that mailed results may take longer to arrive

➢ ***Remember your provider will contact you as soon as they can***

➢ ***If you are not contacted about your results within 48 hours of getting lab work done, feel free to contact your provider***

*** Note: If you receive a radiology exam the bill for reading the exam will come from ABC Imaging.*

(*Continued*)

(Continued)

Reception Phone Call Procedure

Answering Phone Calls Flowchart

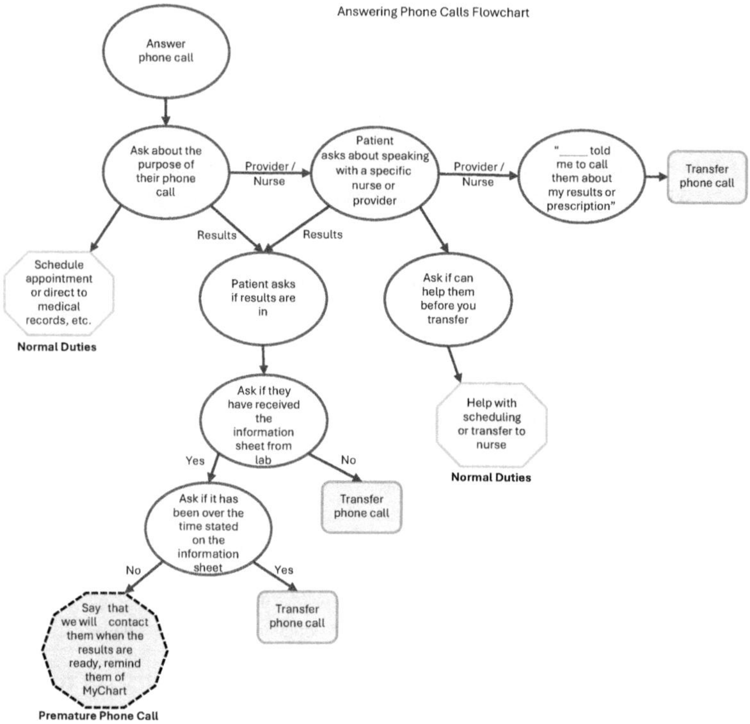

Future State Value Stream Map

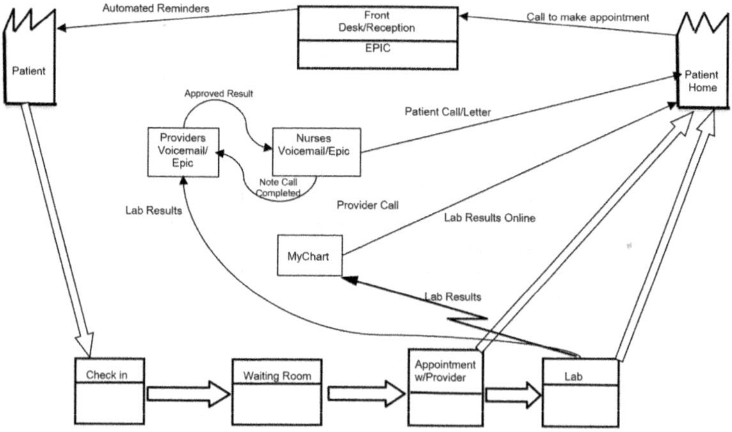

Implementation Plan

Action	Responsible	Deadline	Expected Outcome	Actual Outcome
Reception phone call flowchart. • Draft a flowchart. • Distribute the flowchart after approval.	Rina and Philippe	Monday (week 2)	Reduce the number of calls going to nurses.	Expected outcome met.
Lab to patient communication pamphlet. • Draft three versions. • Discuss with the legal team. • Discuss with the lab team. • Distribute pamphlets to lab and inform them of when to give to a patient.	Rina and Philippe	Monday–Tuesday (week 2)	Reduce the number of calls coming into the clinic and to nurses.	Expected outcome met.
Epic software use training • Observe nurses/providers methods on Epic. • Discuss with IT/Head nurse training. • Establish a standard method. • Train nursing and provider team.	Rina, Sean	Tuesday (week 2)	Increase the speed at which nurses can record and document results and patient contacts.	Expected outcome met.
Time-defined flow of reviewing results and calling patients between/during appointments depending on availability. • Talk to management about possibility of a schedule for managing nurse inboxes. • At the end of the day, check and record numbers for each nurse.	Philippe, Tammy	Wednesday–Thursday (week 2)	Increase the number of calls made to patients within a workday by creating an environment of competition and self-improvement.	Expected outcome met.

Effects Confirmation

Changes Made During Implementation:

1. Talked with nursing staff and standardized documentation.
2. Adjusted language of lab information forms to assure correct directions to log into MyChart were given.
3. Created a tracking sheet for the head nurse to track calls and inboxes of nursing staff.

General Observations:

- Countermeasures were well received by staff.
- People admitted when they sometimes forgot to do countermeasures.
- Unexpected absences of staff within scope occurred.
- Some phone calls regarding lab results could not be made to patients because providers did not release them to their nurse.
- Some data could not be completed, because it hadn't been 24 hours since the patient's appointment.
- Reduced number of patients contacting reception asking for lab results.
- Reduced number of voicemails and phone calls going to nurses.

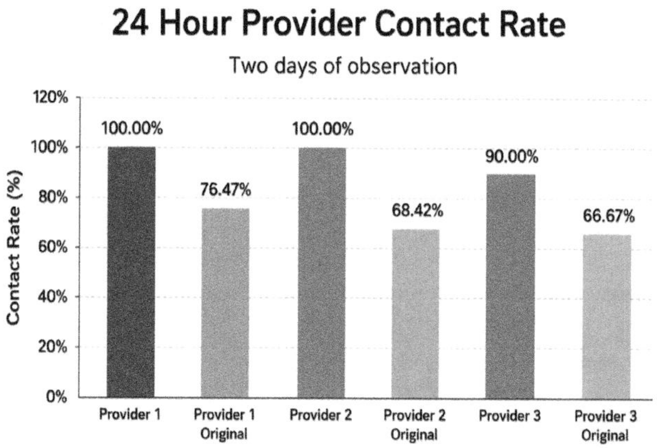

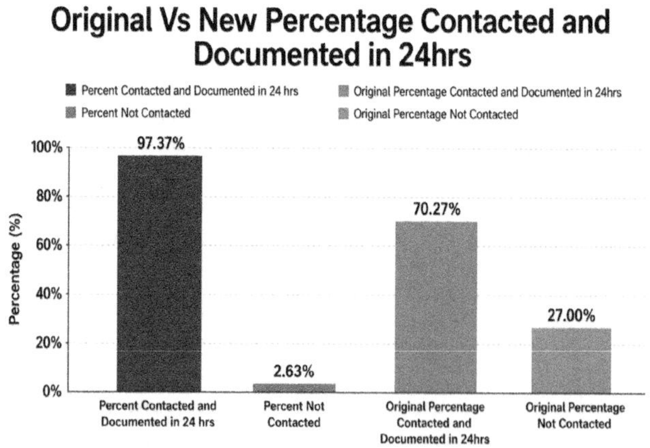

Note: Sample sizes for new data were 38 total ancillary services received collected over two days vs. 74 in the original data.

Goals Met:

- Increased patients, who received ancillary services, being contacted within 24 hours of their lab results becoming available from 70.27% to 97.34%!!!
- Reduced number of patients not contacted with their lab results from 27% to 2.63%!!!

Follow-Up Actions

Action	Responsible Party	To Complete Action	Due
Continue daily recording and checking calls/messages for all nurses.	Tammy	Will continue to record and report daily nurse data.	Ongoing
Give the lab staff note sheet documents to print for future use.	Tammy	Will print and refill pamphlets for the lab to be used.	Ongoing
Monitor progress and ensure changes are still adding value.	Tammy and Philippe	Consult with project owners regarding future changes on a bimonthly basis.	Ongoing
Give the schedule for tallying voicemails and inboxes.	Tammy	Tammy prints out a recording sheet each month.	Ongoing
All nurses continue to use the same procedure for documenting results.	Clinic nurses	Monthly check in that all nurses are following the same procedure to enter results.	Ongoing
Give receptionists the flowchart for future editing.	Clinic Reception	Will print and edit when needed.	1 week

Case Study 5.4: Physical Therapy (PT) Revenue Cycle	Date: XX/XX/20XX
Sponsor: Physical Therapy Clinic	**Author: Alex**

Background

- Their physical therapy (PT) clinic serves a varying population of patients, including children, seniors, high school athletes, and post-surgery inpatients.
- In the past months, the physical therapy clinic has experienced a lack of documentation of patients' visits. This has led to a decrease in revenue, because they cannot charge for an undocumented visit.

Current Condition:

Assumptions: *Inpatients are not being considered; Historical data from the last three months with 1,067 patients seen; Occupational Therapists and Cardiac Rehab are not being considered.*

Current State Value Stream Map (without timeline)

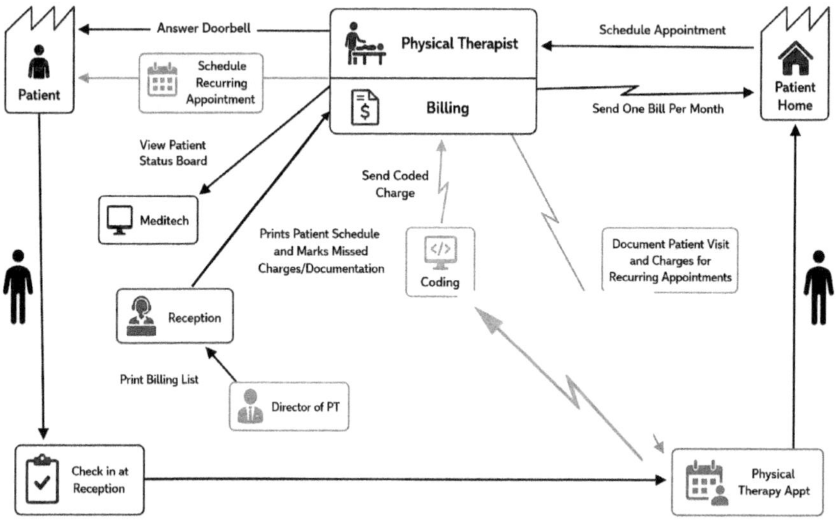

Observations:

- **6.75% of patients seen in PT went uncharged.**
- **Estimated lost charges are $11,866 with a range from $8,303 to $15,620 over 3 months.**
- Appointments go uncharged 0–7 minutes, 8–15 is one charge, 15–22 is one charge, 23–30 is two charges, etc.
- $0 charge visit documentation.
- One PT is scheduled on "the floor" each week to complete PT rounds with inpatients.
- Documentation of a patient encounter needs to occur within 5 days of the patient's appointment.

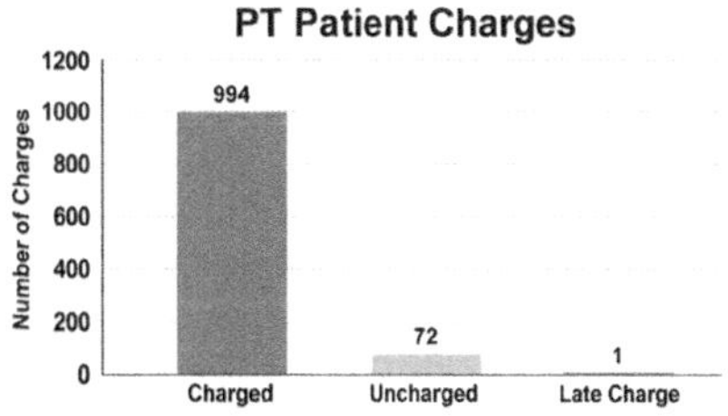

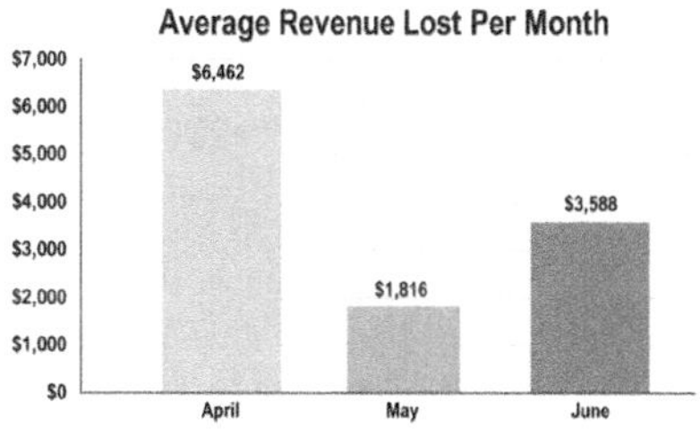

Major Problem: No documentation of patient visits, so charges cannot be sent to insurance/patients for billing.

SMART Goal:	• Decrease the number of uncharged patient encounters by 3% (from 6.75% to less than 3.75%).

Root Cause Analysis

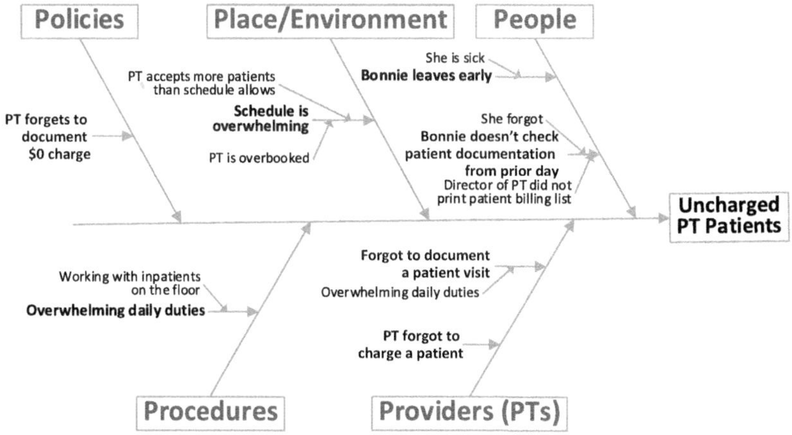

5 Whys:

Problem Statement 1: PT patient goes uncharged for their visit.

↳ **Why?** A bill was never sent to insurance or the patient.

↳ **Why?** Charges were never added to patient documentation.

↳ **Why?** PT forgot to add charges to documentation.

↳ **Why?** PT got distracted from documenting.

↳ **Why?** PT had to go answer the doorbell to let a patient in.

Problem Statement 2: PT patient goes uncharged for their visit.

↳ **Why?** A bill was never sent to insurance or the patient.

↳ **Why? Patient** documentation was never filled out by PT.

↳ **Why?** PT didn't have time to fill out the documentation.

↳ **Why?** They were on the floor working with inpatients and had their regular appointments.

Problem Statement 3: PT patient goes uncharged for their visit.

↳ **Why?** A bill was never sent to insurance or the patient.

↳ **Why?** Patient documentation was never filled out by PT.

↳ **Why?** PT was too busy to fill out the documentation.

↳ **Why?** The PT schedule did not allow time to document patients before the next patient appointment.

Problem Statement 4: PT patient goes uncharged for their visit.

↳ **Why?** A bill was never sent to insurance or the patient.
↳ **Why?** Patient documentation was never filled out by PT.
↳ **Why?** PT was too busy to fill out the documentation.
↳ **Why?** The PT was busy covering administrative duties.
↳ **Why?** Reception left early and there is no coverage.

Problem Statement 5: PT patient goes uncharged for their visit.

↳ **Why?** A bill was never sent to insurance or the patient.
↳ **Why?** Patient documentation was never filled out by PT.
↳ **Why?** PT forgot to fill out documentation for the patient.

Problem Statement 6: PT patient is not properly charged for their visit.

↳ **Why?** PT does not charge for every treatment performed at the visit.
↳ **Why?** PT does not properly document the patient visit.
↳ **Why?** PT documented significantly after the patient appointment.
↳ **Why?** PT did not have time after the patient appointment to document.
↳ **Why?** PT had another patient scheduled immediately after the first appointment.

Problem Statement 7: Insurance will not pay for a patient's surgery.

↳ **Why?** Lack of documentation, proving that the patient went through PT.
↳ **Why?** PT never filled out the proper documentation.
↳ **Why?** They were too busy/forgot to fill out documentation.
↳ **Why?** Something in the current system is not working.

Proposed Countermeasures

- 5-minute scheduled time between appointments for documentation.
- Reduce Director responsibilities/delegate administrative tasks.
- Electronic reminder system in Meditech.
- Print patient list and give to PTs at the beginning of the day.
- Bring Surface Pro in during the patient appointment to document.

Target Condition:

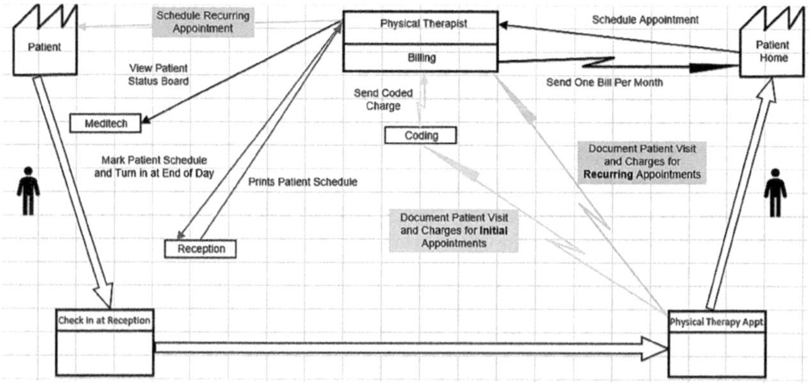

Implementation Plan:

Action	Responsible	Deadline	Expected Outcome	Actual Outcome
Schedule patients' appointments with 5 minutes between appointments.	Bonnie and Alex	Monday (week 2)	More accurate documentation. Increase in revenue.	Not possible as all appointments for the week were already scheduled.
Determine tasks that can be distributed.	Colin and Alex	Monday (week 2)	Allow for Director to document more frequently.	Expected outcome met.
Create reminder in Meditech for undocumented patients for each PT.	IT and Alex	Monday (week 2)	Increase documentation. Eliminate need for Bonnie and Colin to print patient sheets.	Not possible with current technology or level of training.
Allow other PTs to answer phone calls.	Colin, Bonnie, and Alex	Tuesday (week 2)	Allow for Director to document more frequently.	Expected outcome met.
Have Bonnie print each PT their patient list for the day.	Bonnie and Alex	Tuesday (week 2)	Increase documentation. Decrease late charges.	Expected outcome met.
Have each PT mark the patient when they are done documenting.	PTs	Tuesday (week 2)	Increase documentation. Decrease late charges.	Expected outcome met.
Each PT will hand in a finished sheet at the end of the day.	PTs and Bonnie	Tuesday (week 2)	Increase documentation. Decrease late charges.	Expected outcome met.
Each PT brings Surface Pro into patient appointments.	PTs	Tuesday (week 2)	Document patients during appointment. More accurate documentation charges. Increase revenue.	Some PTs were not willing to bring Surface Pro into appointments to document.

Effects Confirmation

Data collected for two days of observation after implementation of countermeasure, with 44 patient appointments observed.

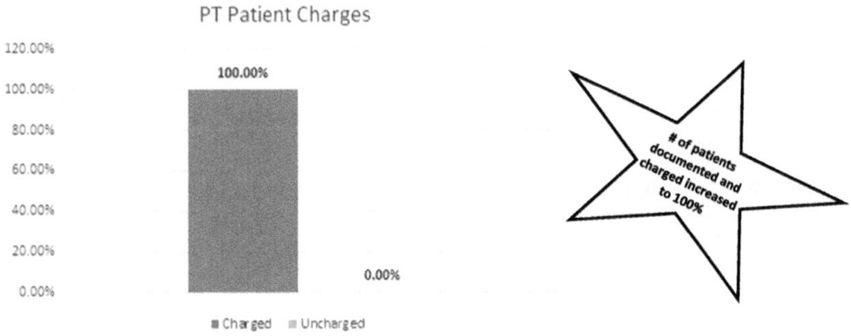

Goal Met: Decreased the number of uncharged patient encounters from 6.75% to 0.0%!!!

Follow-Up Actions

Action	Responsible Party	To Complete Action	Due
Doorbell should be eliminated because it is an unnecessary workflow interruption.	Danielle and Colin	Find out why the doorbell cannot be eliminated. Ask "why not?"	In 3 weeks
Have Colin delegate tasks to other PTs.	Colin and PTs	Total joining meeting to Earl, transitional care meeting to Fiona. For inventory management and ordering Fiona will be responsible for ordering needles and Gaston will order tape; also each PT will have the ability to order what they need for treatments.	Next 4 weeks

Case Study 5.5: Patient Throughput at a Clinic	Date: XX/XX/20XX
Sponsor: Clinic	**Author: Chris**

Background

- The clinic is experiencing long wait times from check in to a patient's contact with the provider and is seeking to reduce wait times and subsequently increase patient flow.
- The clinic sees 66 patients a day on average across six providers.
- Each portion of the clinic, including family practice (FP), Physical Therapy (PT), and Internal Medicine (IM) rotates taking clinic phone calls.
- Patient appointment times vary by provider from 10 minutes for walk-in appointments to 60 minutes for annual physicals.

Current Condition:

Assumptions:

- *Data was collected from days during a normal week of clinic operations on four consecutive days for week 1.*
- *It is assumed Family Practice providers saw a normal volume of patients.*
- *Check in and Check out times based on timestamp recorded when the provider, nurse, or receptionist logged into Epic.*

Current State Value Stream Map (without Timeline)

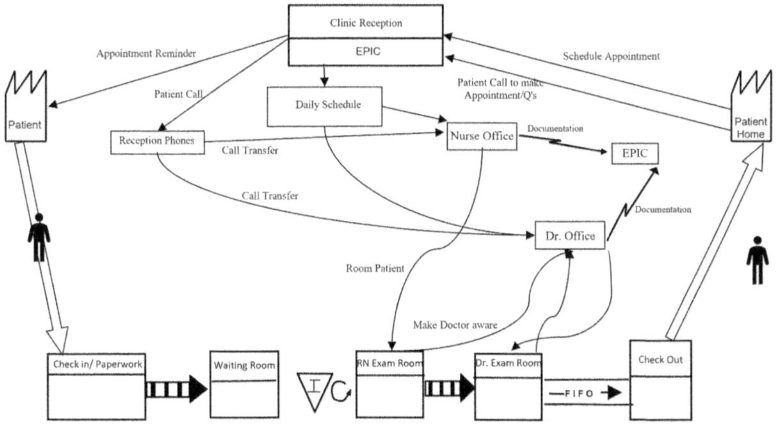

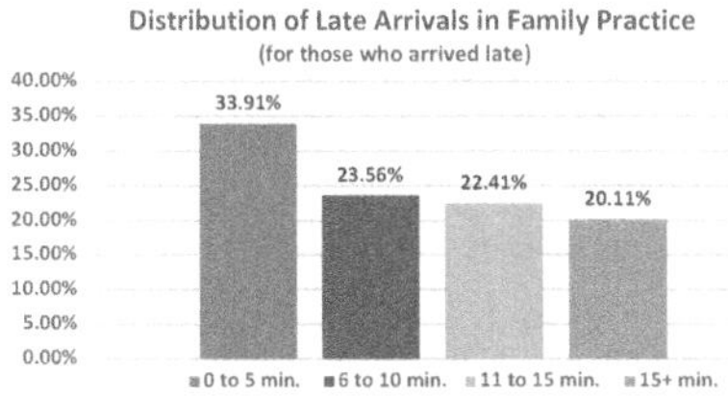

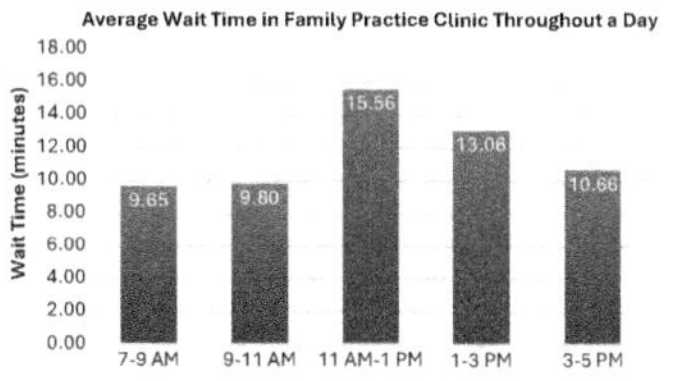

Key Stats:

- 48.79% of patients arrived late to their appointments' expected arrival time.
- On average, patients spend approximately 12.55 minutes waiting during a visit.
- Standard Deviation of Average Daily Wait time of 4.17 minutes.
- Approximately 42.2% of patients were seen by a provider in 5 minutes or less.

Note: Average wait time calculated assumes the following

- *Waiting only begins at or after the appointment start time.*
- *For patients who arrived at or early to their expected arrival time, only time from room entry to provider visit was counted as wait time.*

Major Problems:

- Front desk staff are overwhelmed with provider specific duties and phone calls causing longer check in and wait times.
- Local patients expect personalized and in depth care, results in expectation to help with all things and answer all questions regardless of appointment length, or staffing.
- Appointment times are not representative of actual time spent with patients.
- Patients arriving late is causing back up of appointments, creating longer appointment times and subsequently longer waits that build through peak hours.

SMART Goals:	• **Decrease percentage of patients who arrive late 15 minutes or more from 20.11% to 10% or less.** • **Decrease total average wait time to 10 minutes or less.**

Root Cause Analysis:

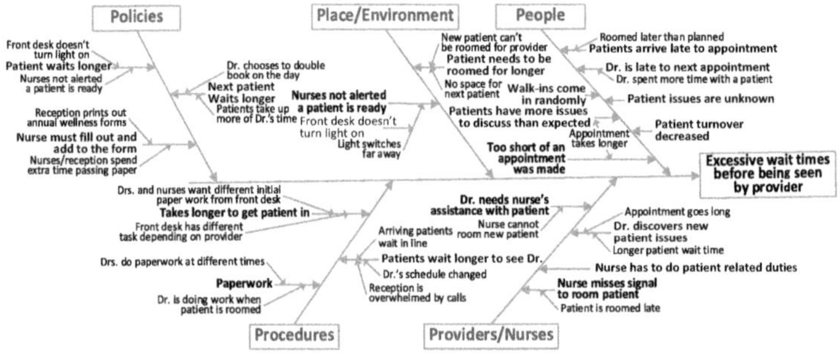

5 Whys:

Problem Statement 1: Excessive patient wait times.

↳ **Why?** Lines form at Family Practice check in.
↳ **Why?** Front desk staff are overwhelmed with admin tasks.
↳ **Why?** An excess of phone calls come to the front desk during peak hours.
↳ **Why?** Patient expectations set by staff to fulfill even the smallest needs are extremely high.
↳ **Why?** Extremely patient centered culture.

Problem Statement 2: Excessive patient wait times

↳ **Why?** Dr.'s appointments go past appointment end time.
↳ **Why?** Previous appointment went long.
↳ **Why?** Dr. spends extra time with a patient to assure needs are met.
↳ **Why?** The patient arrives late to expected appointment time.
↳ **Why?** Methods of patient reminders are ineffective.

Problem Statement 3: Excessive patient wait times.

↳ **Why?** Dr.'s appointments go past set appointment end time.
↳ **Why?** Previous appointment went long.
↳ **Why?** Dr. spends extra time with a patient to assure needs are met.
↳ **Why?** Appointment length was not fitting to needs of the patient.
↳ **Why?** The patient came in with more to discuss than expected.

Problem Statement 4: Excessive patient wait times

↳ **Why?** The patient sits in the waiting room too long.
↳ **Why?** The nurse comes out late to room patients.
↳ **Why?** The nurse continues to do other essential tasks.
↳ **Why?** The nurse does not see light to alert them that a patient is ready.
↳ **Why?** Front desk staff does not turn on alerting light.
↳ **Why?** Light controls are in an inconvenient location for busy staff to use.

Problem Statement 5: Excessive patient wait times.

↳ **Why?** Patients are roomed later than expected.

↳ **Why?** Drs. and nurses spend extra time with the preceding patient.

↳ **Why?** Dr. does not see the patient soon after the nurse rooms the patient.

↳ **Why?** Dr. does paper work between appointments.

↳ **Why?** Drs. have different methods of best practice.

Proposed Countermeasures

Counter Measures	Needed Actions	Responsible	Expected Outcome
1. Standardize the patient arrival alert system across providers.	• Discuss the best method with nurses and providers.	Chris	Decrease wait time by alerting staff that the patient is ready as soon as possible.
	• Create reminders for front desk staff to turn on the light.	Chris	
	• Pilot with front desk staff.	Chris	
2. Patient expectation and wait time pamphlet.	• Create pamphlets.	Chris	Decrease incoming calls and decrease the amount of patients arriving late by resetting patient expectations and decreasing the level of communication expected of providers and nurses.
	• Modify and edit with feedback from providers and admin staff.	Chris, Kirk, Lisa, Debby, all providers	
	• Establish location of hand-out during appointment.	Chris, providers	
	• Have Dr.'s hand-out pamphlet at end of each appointment.	Providers, front desk staff	
3. Change how rooming paperwork is sorted and used by nursing staff.	• Discuss with providers, admin.	Chris, nurses	Would allow for decreased wait times by removing the need for front desk to sort paperwork.
	• Modify when sheets are signed, moved sheets to nursing cubes.	Chris, nurses	
	• Pilot new method.	Chris, nurses	
4. Have all doctors and nurses reschedule nonessential patients if they are late by greater than 15 minutes.	• Discuss with providers, nurses, admin.	Admin, Chris, Kirk, providers	Decrease patient wait time and allow more patients to be seen and doctors more time to deal with daily variation.
	• Make changes to timing, implement a call system to tell patients of new standard, post notice in all rooms.	Chris, providers	

Target Condition:

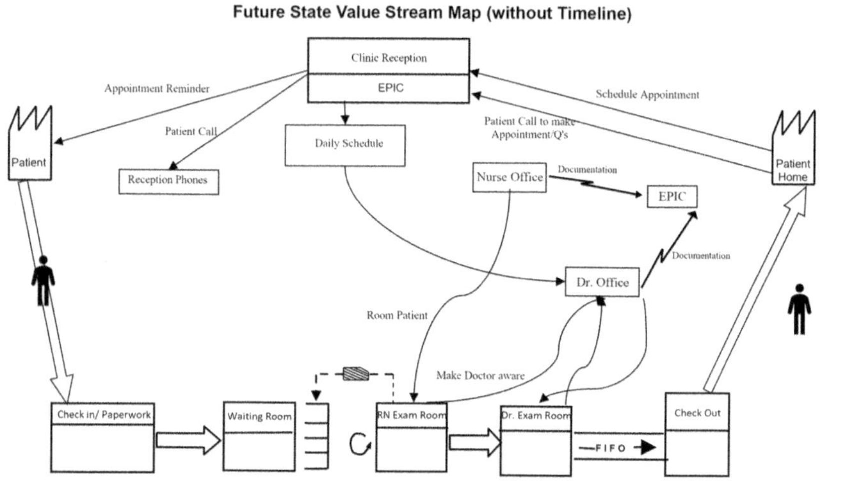

Implementation Plan:

Action	Responsible	Deadline	Expected Outcome	Actual Outcome
Create a reminder system for front desk to turn on lights to alert nurses of patient arrival.	Chris	Monday (week 2)	See Countermeasure #1.	Expected Outcomes met.
Give a pamphlet to Dr. and place in offices or front desk to hand out at the end of appointments.	Chris, Providers	Monday (week 2)	See Countermeasure #2.	Expected Outcomes met.
Discuss with staff a 15 minute late turn away policy for all patients that come in 15 minutes or later.	Chris, Kirk, nurses, Providers.	Monday (week 2)	See Countermeasure #3.	Expected Outcomes met.
Standardize patient info paper, and printed patient forms used by select staff. Move location of forms, and when they are filled out by staff.	Chris, select Providers	Tuesday (week 2)	See Countermeasure #1.	Expected Outcomes met.

(Continued)

(Continued)

Action	Responsible	Deadline	Expected Outcome	Actual Outcome
Create expectation and appointment reminder pamphlet for patients.	Chris	Tuesday (week 2)	See Countermeasure #2.	Expected Outcomes met.
Implement rescheduling policy in pilot study.	Chris, Kirk, Providers	Wednesday (week 2)	See Countermeasure #4.	Expected Outcomes met.

Effects Confirmation:

Changes Made During Implementation:

1. Changed wording on the pamphlet to be more appropriate to purpose.
2. Staff requested in room versions of pamphlet to reference for patients.
3. Adjusted reminders for front desk to better fit providers' and nurses' needs.
4. Created notice to alert patients of new rescheduling policy and posted across clinic.

General Observations:

- Changes made were implemented smoothly with staff assistance.
- Shortages in staffing were observed during the implementation period.
- Providers noted the amount of time late to qualify for rescheduling could be shortened significantly.

Note: Average wait time calculated assumes the following over the 3-day implementation period:

1. *Waiting only begins at or after the appointment start time.*
2. *For patients who arrived at or early to their expected arrival time, only time from room entry to provider visit start was counted as wait time.*

Key Stats:

- As before, approximately 50% of the patients arrived late to their appointments' expected arrival time.
- On average, patients spend 10.05 minutes waiting during a visit.
- Standard Deviation of average daily wait time from 4.17 to 2.26 minutes.
- Approximately 44% of patients were seen by a provider in 5 minutes or less.

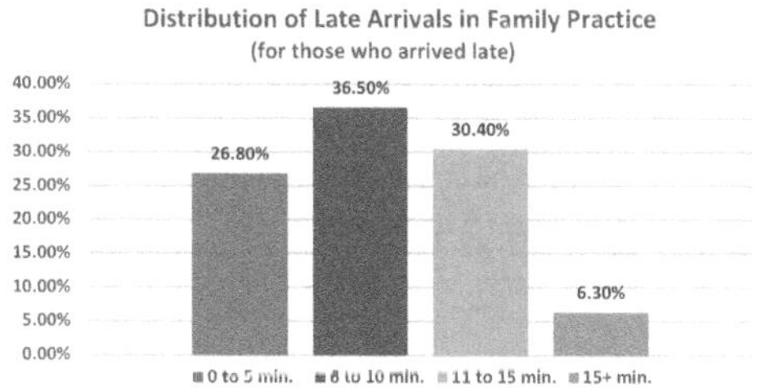

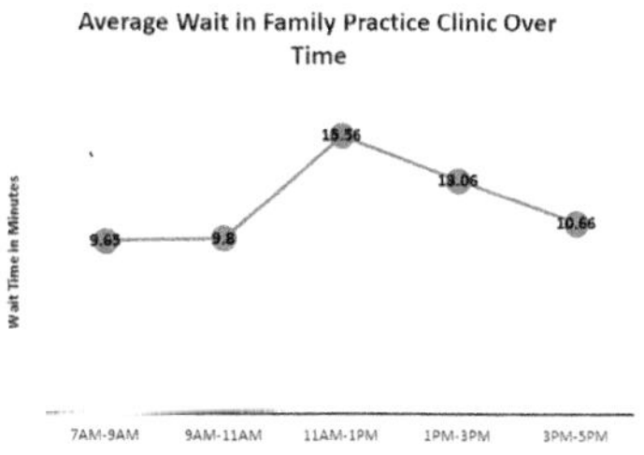

Goals Met:

- **Decreased average wait time from 12.5 minutes to approximately 10 minutes!**
    - Also decreased standard deviation from 4.17 to 2.26 minutes.
- **Decrease the percentage of patients who arrive more than 15 minutes late from 20.11% to 6.30%.**
    - *Note: The percentage of 0–5 minutes decreased while the percentage of 6–10 and 11–15 minutes increased. The system should be adjusted by rescheduling patients who are 5 or 10 minutes late.*

Follow-Up Actions

Action	Responsible Party	To Complete Action	Due
Reduce rescheduling time limit to any time beyond appointments scheduled start time to 5 minutes.	Kirk, Leslie, IM, and FP providers	Send out a new message to patients, post announcements in the IM wait room.	In 1 week.
Create standard buffer in time between appointments, depending on their length to allow doctors to stay on schedule.	Kirk, IM, and FP providers	Implement a time buffer between appointments based on provider feedback. Pilot new schedule first before full implementation, introduce slowly to ease use.	In 3 weeks.
Implement patient expectation pamphlets in internal medicine and create a version in Epic to attach to post appointment summaries.	Kirk, Leslie, IM providers	Teach nurses and doctors in internal medicine how pamphlets should be distributed.	In 3 weeks.
Pilot a study using the Vocera communication system between nurses and front desk, and test communication effectiveness.	Maintenance	Have maintenance and IT coordinate on training nurses on system and trial for a week.	In 3 weeks.
Modify appointment lengths to better suit patient needs, specifically walk-in appointments.	Kirk, IM, and FP providers	Work with staff to establish a standard appointment length for each type of visit that accounts for extra time spent with patients.	In 3 weeks.

6 Medical, Surgical, and Emergency Departments Workflow Case Studies

In this chapter, the RIE case examples contain a wide array of common workflow issues found in the key departments of hospitals. The examples illustrate how issues often arise due to multiple interacting factors and therefore require multiple countermeasures to realize an overall improvement. Making workflows visible is essential to redesigning them to be more efficient and effective.

The primary objective in Case 6.1 is to reduce the turnaround time on lab orders from a hospital's emergency department (ED). This RIE brief uses a swimlane chart, which is a special kind of flow map that separates the workflows of different actors or departments into "lanes" to illustrate the coordination of activities among them. By addressing several underlying causes with straightforward, low-cost solutions, the team was able to pilot the new workflows and achieve significant reductions in turnaround time. The team identified several ways to further improve the workflows, which are listed in the Follow-up Actions table at the end of the RIE brief.

In Case 6.2, the RIE team improves the workflow in the Sterile Processing Department (SPD) through an improved department layout. This RIE brief features a spaghetti diagram that illustrates the movement of personnel and/or materials on the layout drawing of the space. The new layout supports the order of activity decrease in the workflow, resulting in a 49% decrease in the walking distances required of SPD technicians, a 60% decrease in the time spent on the non-value-adding activities of walking and searching for materials, and a 10% increase in usable space (critical for cart maneuvering).

Case 6.3 utilizes a mock trauma event to study the workflows in a small ED. Analysis of the workflows reveals improvement opportunities in both the emergency room (ER) layout and in the management/organization of medical supplies. A second mock trauma event demonstrates a 46% reduction in search time, a 73% reduction in the number of times staff members had to maneuver around one another (called "interferences" in the workflow), and high staff satisfaction.

Case 6.4 uncovers a host of interrelated factors influencing the barcode scanning rate for patient encounters and medication deliveries for inpatients. In the current state, 15% of patient encounters and 26% of medications were not entered into the electronic health record system via barcode, introducing risk of inaccurate or missed information in the Electronic Health Record (EHR). Careful study of the workflow found opportunities to improve the computer workstation layout, barcoding procedures, and computer equipment. Preliminary testing saw virtual elimination of

DOI: 10.1201/9781003670698-6

missed scans, though additional confirmation was needed beyond the RIE testing day.

The final case study in this chapter deals with intravenous (IV) medications delivered to the medical floor's automated medication storage system called an Omnicell. The hospital pharmacy technicians deliver IV bags to the refrigerated unit of the Omnicells located on the floor according to physicians' orders. When nurses need to administer an IV medication, they frequently had trouble locating it. Changes to the workflow included standardizing and labeling the locations of IV medications, creating a visual signal on the refrigerated unit that indicated that the medication had been delivered, and a specified protocol in the event a nurse did not find the medication. Results were generally positive but limited to day shift, in large part because the night shift stakeholders were overlooked, which was a key learning by the RIE team.

As you review the cases, you will likely note how workstation and workplace layout can either help or hinder workflow. Significant improvements can be made by organizing workspaces to support the workflows of the individuals working there. You will also note how often standardization is a powerful weapon in the arsenal to defeat inconsistent, inefficient, and ineffective workflows.

Case Study 6.1: Increasing the Reliability of Lab Orders Placed by the Emergency Department (ED)	Date: XX/XX/20XX
Sponsor: ED and Lab	Author: Paula

Background

- ED lab orders are given priority.
- The hospital is in process of new software upgrade. Planned Implementation in December.
- Blood samples are kept in a fridge for 1 week.
- Urine samples are kept for 24 hours.
- Waiting orders are attached to a magnet clip on a refrigerator.
- Extra samples without orders have the patient's name written on a dry-erase board on the refrigerator.

Current Condition

Swimlane Diagram:

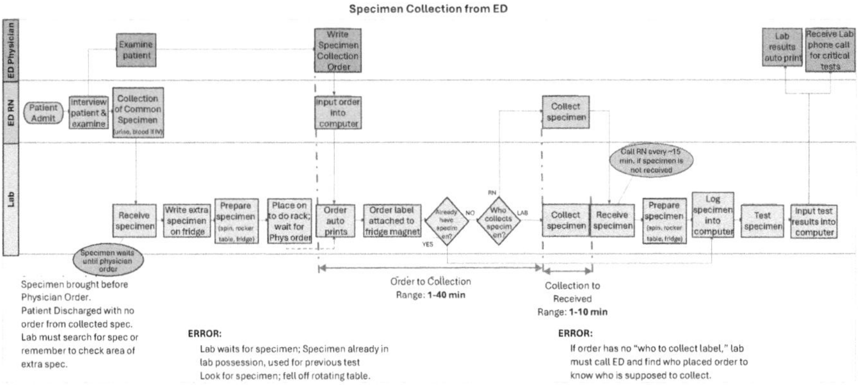

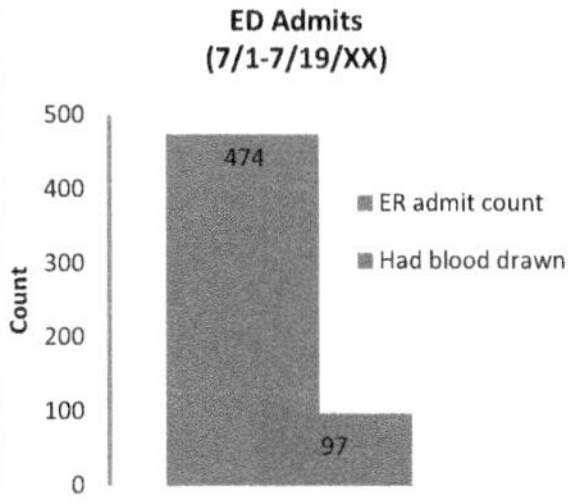

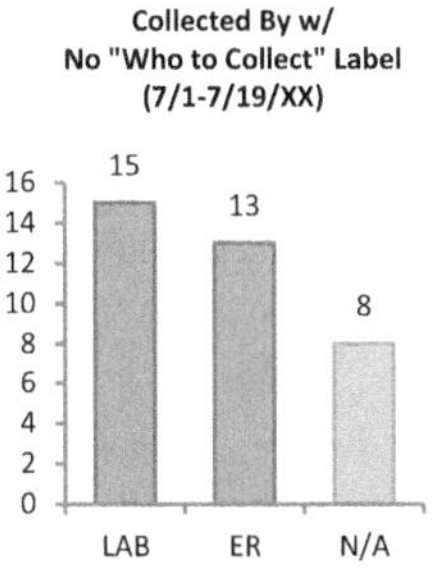

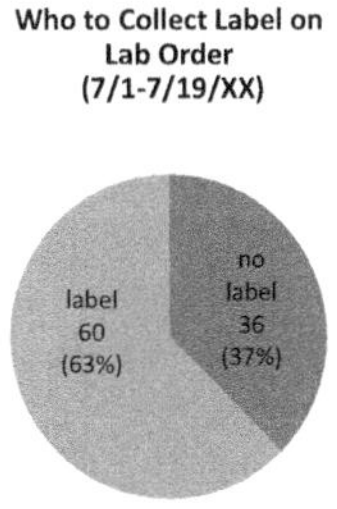

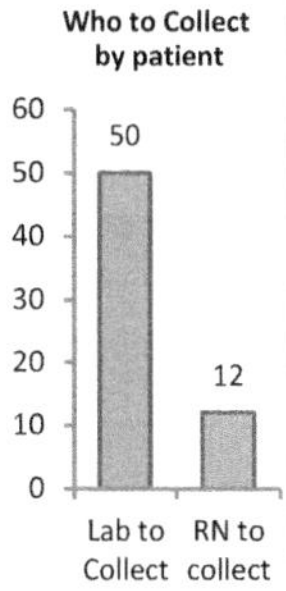

(patients)	avg/day	std dev
ER admit count	24.9	5.7
Had blood drawn	5.1	2.6

Process Times		
(minutes)	LAB	ER
ORD to COLL	22.75	-13.93
COLL to RECV	7.75	37.40
ORD TO RECV	30.50	23.47

47 - Patients
8–specs collected before order by ED

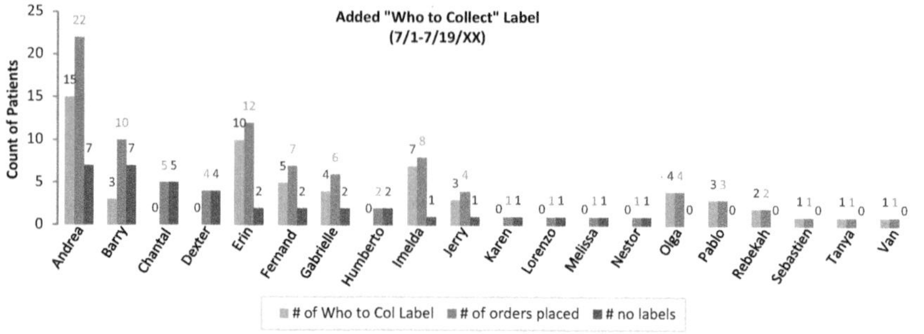

Communication Methods:
Order Labels
Phone Calls
In Person

Label vs No Label		
(minutes)	AVG	STD DEV
LAB NO LABL	28.6	24.6
LAB LABL	17.3	10.4
ER NO LABL	20.3	12.9
ER LABL	11.3	6.3

39.5% faster
w/ label

44.3% faster
w/ label

SMART Goal:	• Reduce cycle time from the physician's order of specimen to lab receiving it in under 25 minutes.

Root Cause Analysis

1. **Lab making phone calls to ED.**
   a. Confusion on who to Collect.
      i. No Who to collect label.
         1. Inexperienced about computer order input.
         2. No requirement in software.
         3. Registered Nurse (RN) collect only specimen.
   b. Waiting to receive specimen.
   c. Confusion about order.
2. **ER Collect Before Phys Order.**
   a. Convenience of collect with IV start.
   b. Have the sample before introducing other factors into body (pain meds, chemicals from IV).
   c. High probability of future collects.
3. **Takes longer than 25 minutes.**
   a. Out of needed supplies.
   b. Busy with other tasks.
   c. Waiting for other instructions.
   d. Forgot to send to lab.
4. **Waiting for Specimen.**
   a. Not received in lab.
   b. Already in lab, but unaware of possession.
   c. Not written on the fridge.

Proposed Countermeasures

1. **Reduce Phone Calls to Lab.**
   a. Mandatory Lab to collect, registered nurse (RN) to collect, or Add On order label with each physician order.
2. **Remove decision steps of the specimen collection process.**
   a. Discontinue collecting specimen because of convenience.
      i. Only collect when labs are ordered.
      ii. Also remove the step of lab checking to see if order is already in lab.
   b. Only have lab collect blood specimens from the patient.
      i. Reduces confusion of who is to collect specimen.

Target Condition for Countermeasure #2:

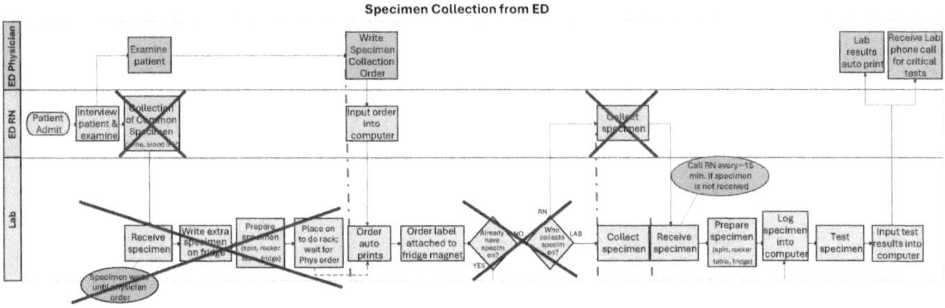

3. **Create Spec Status Board for in Lab.**
   a. Place Col Magnet when Order printed.
   b. Place a Receive magnet when spec brought into lab.
   c. Extra if Registered Nurse (RN) collects before Phys order or extra test spec.

Target Condition for Countermeasure #3:

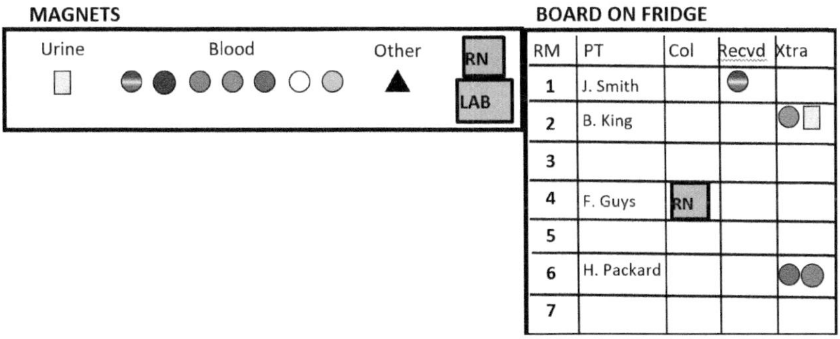

4. **Drop area of ED collected specs.**
   a. Place close to door to reduce travel into lab.
   b. Have signal light or bell to notify lab of spec received.

Implementation Plan

Action	Responsible	Deadline	Expected Outcome	Actual Outcome
Develop plan to implement.	Paula, ED, Lab, Tobias	Friday (week 1)	Buy-in from stakeholders.	Unable to receive buy into plan of removing collect before order and only lab to collect; received buy-in Monday week 2 with labeling each order, fridge magnet board in lab, and moving drop bin to ED with signal switch.
Train ED about labeling orders.	Paula, Shary	Monday (week 2)	Train ED staff.	Trained ED staff on floor and made a reminder note to stick on the computer monitor to make labels for each order; developing order to state "sample in lab."
Find signal switch to place in ED to signal lab.	Paula	Monday (week 2)	Install a wireless signaling device.	Found wireless switch; signal does not reach to lab. Of the 100 ft listed range, only works about 30 ft in line of sight.
Create a magnet board for the fridge.	Paula	Monday (week 2)	Magnet board installed.	Status board and procedures made and posted.
Write procedures for labeling, magnet fridge, and signal switch.	Paula	Tuesday (week 2)	Completed procedures.	Created and revising w/ restrictions found.
Pilot run of label for each order, magnet board, and signal switch.	Paula, ED, Lab	Monday–Tuesday (week 2)	Run pilot.	ADD ON Label programmed around noon on Tuesday; signal switch remained in lab; label instruction on each computer in ED.
Review plan.	Paula, Tobias, Shary, Owen	Wednesday (week 2)	Review plan.	Talked with Owen, Shary, and Tobias about current implementation and made revisions on.
Communicate results.	Paula, ED, Lab, Tobias	Thursday (week 2)		Results and fine-tuned process communicated.

Effects Confirmation

1. **Problem:** Lab phone call to ED for direction about spec order

   Pilot Implementation:
   - Inform ED to input instruction label with each order.
   - Created "Add On" label for specs already in lab to test.

Ordered to Received Times

	AVG (min.)	STD DEV (min.)	COUNT	#>25 min	PREV AVG	% Decrease
ER	14.67	12.51	6	1	23.47	38%
LAB	10.5	1.5	2	0	30.5	66%

ER	Avg (min.)	Std (min.)	Count
with label	10.00	7.56	5
No Label	38.00	0	1

2. **Problem**:
   - Lab unaware of what specs have been collected for the patient.
   - Lab waiting for spec when the test was added on order.

   Pilot Implementation:
   - Patient status board.
   - "Add On" order instructions if spec has been sent to lab.
   - Makes Lab aware of what types of specs have been collected.

Patient Spec Status Board in Lab

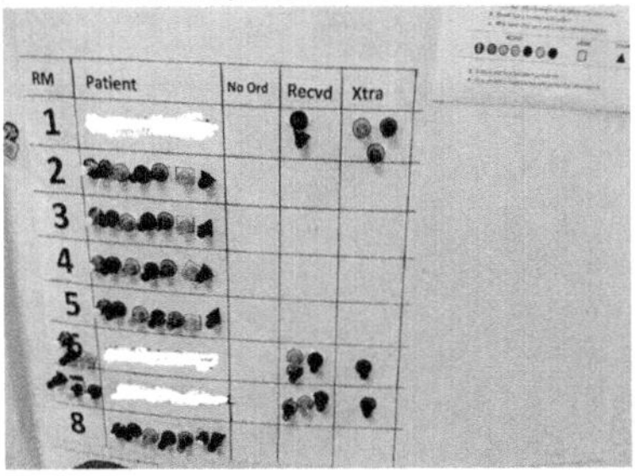

Signal Switch at Spec Drop Off

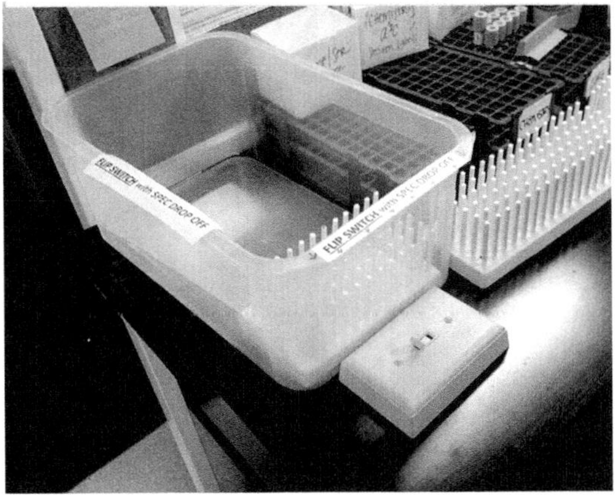

Follow-Up Actions

Action	Responsible Party	To Complete Action	Due
Find better method of signal lights. Switch for multiple lights throughout lab. Improve reliability with wired vs wireless. Blinking light.	Owen	-	In 1 week.
Improved convenient location for spec drop between ED and lab. Closer to door. In ED. Drop window. Door slot.	Owen, Shary	-	In 1 week.
Bring up order labeling to ED staff meeting. Lab needs instructions to reduce phone calls to ED. Decreases confusion.	Shary	-	In 2 weeks.
Bring up spec status board in staff meet.	Owen	-	In 2 weeks.

Case Study 6.2: Improve the Flow of Traffic and Create a More Efficient Work Environment in the SPD	Date: XX/XX/20XX
Sponsor: ED	Authors: Ana and Bill

Background

- The Sterile Processing Department (SPD) provides sterilization services to all of the hospital system including the seven operating rooms, ER, and all clinics. In addition, they service over 30 additional facilities.
- The current layout of SPD does not provide adequate or safe storage for the carts used to transport all of the medical supplies, both before and after sterilization.
- The current location of the workstations in SPD requires excess travel by staff in order to complete the assembly of the kits.
- A lack of organization slows process times and leads to overall inefficiency.

Current Condition

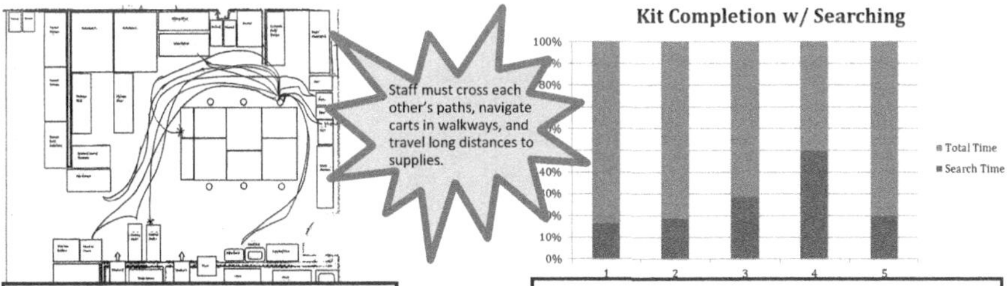

Spaghetti Diagram showing the current flow of work through the space. Current average distance traveled while completing a kit is approximately 232 feet.

Percent of time spent traveling and locating items for kit completion averages approximately 26% of the total time.

Current Layout

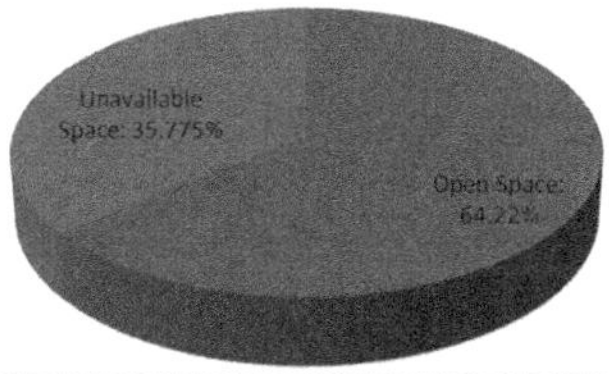

Open space includes all accessible areas used by staff and carts. Unavailable space is being used by machines and fixed storage units.

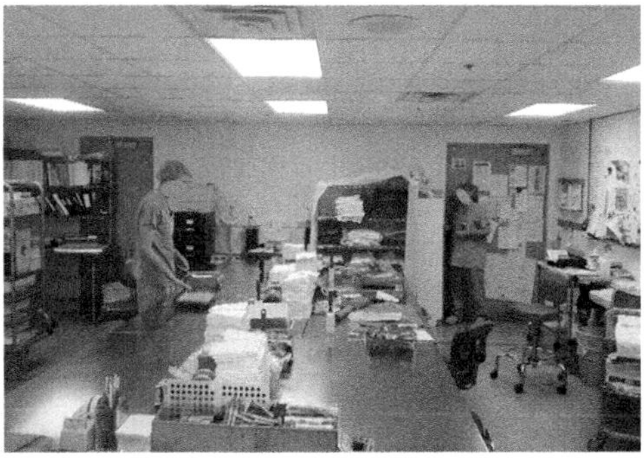

Major Problems

- Excess distance is traveled by workers while assembling kits.
- Kit assembly times are increased due to travel and searching.
- Not enough floor space is available for staff and the storage of carts.

SMART Goals:	
	• Decrease the distance traveled by staff while performing their duties by at least 30%. • Increase available floor space for easier storage of carts and a safer work environment by at least 10%. • Create a more efficient space, thereby decreasing time spent traveling and locating items by at least 10%.

Root Cause Analysis

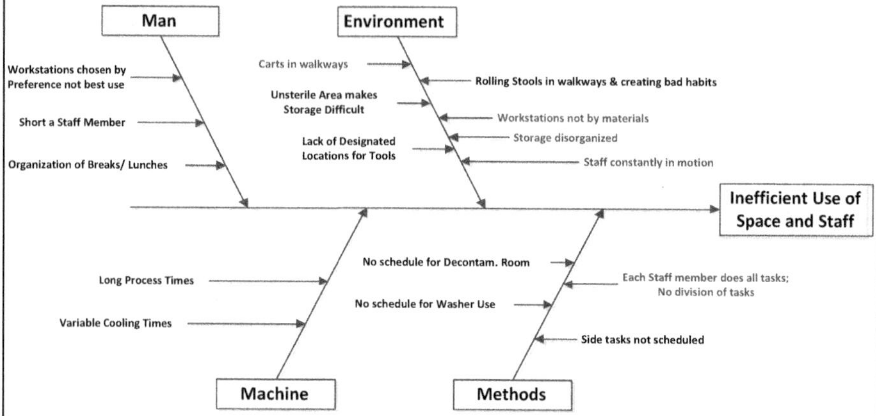

5 Whys Analysis: Inefficient use of space and staff.

Problem 1: Carts in Walkways.

↳ **Why?** No other place to put carts.

↳ **Why?** Not enough room in the current layout.

↳ **Why?** Unorganized workspace and storage space.

↳ **Why?** No designated locations for things to be placed.

Problem 2: Staff Constantly in Motion.

↳ **Why?** All staff members are multitasking.

↳ **Why?** Everyone is expected to do everything, all the time.

↳ **Why?** No one is delegating responsibilities.

↳ **Why?** Work does not have a set order to be performed.

Problem 3: Storage Disorganized

↳ **Why?** Stuff being stored in improper storage units.

↳ **Why?** No one wants to get rid of the storage units.

↳ **Why?** Staff doesn't know if the units will actually be replaced, and they need the storage.

↳ **Why?** They are not asking for the new storage units.

↳ **Why?** Poor communication between departments.

Proposed Countermeasures

1. Move the rack in the decontamination room to the wall.
2. Organize back storage hallway.
3. Switch wrap holder and linen storage with files, computer, and testing equipment.
4. Re-configure workstations.
5. Move decontamination supplies into a closet.
6. Prioritize Cooling Racks by washers.
7. Designate area for the unsterile autoclave cart to be loaded.

Target Condition:

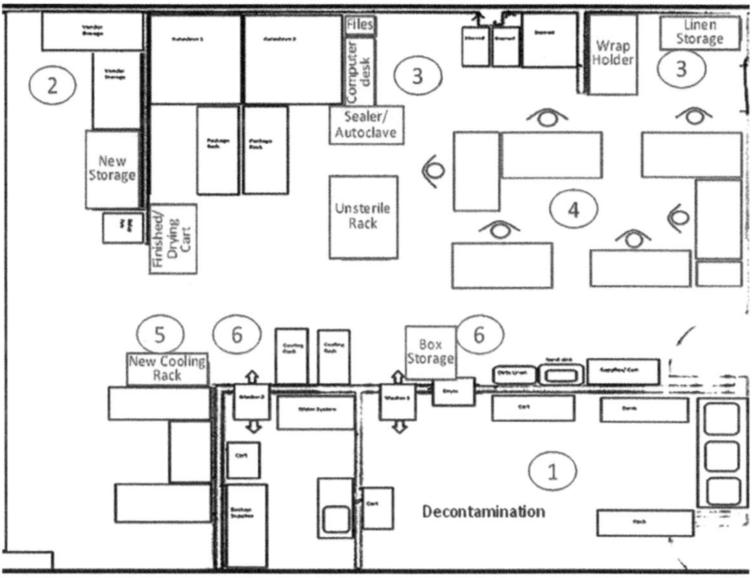

Implementation Plan

Action	Responsible	Deadline	Expected Outcome	Actual Outcome
Place work order to move the rack in the decontamination room to the wall.	Claudette	Friday (week 1)	Less cart traffic/more working space.	Expected outcome met.
Prioritize cooling racks by washers.	Ana, Bill	Monday (week 2)	Reduce search times/ improve task identification by staff.	Expected outcome met.
Designate area for the unsterile autoclave cart to be loaded.	Ana, Bill	Monday (week 2)	Reduce material handling and create more open space.	Expected outcome met.
Re-configure to the first workstation layout.	Ana, Bill	Monday (week 2)	Increase available floor space/improve overall flow through work areas.	Expected outcome met.
Organize back storage hallway by: • Removing old storage racks. • Moving the extra wrap rack and adding a top shelf. • Organizing and labeling.	Claudette, Ana, Bill, Danny	Monday–Wednesday (week 2)	Reduce search times/ and distance traveled.	Expected outcome met.
Switch wrap holder and linen storage with files, computer, and testing equipment. Data/phone cords that reach. Obtain a top shelf for linen. Sort through the "Random" cart. Remove tape.	Ana, Bill, Elsa, Ana, Entire Staff	Monday–Wednesday (week 2)	Reduce distance traveled/use other door for deliveries and outside departments.	Expected outcome met.

(*Continued*)

(Continued)

Action	Responsible	Deadline	Expected Outcome	Actual Outcome
Observe and receive feedback from staff about the layout.	Elsa, Fred, Bill and Ana	Tuesday (week 2)	Improve the first layout.	Expected outcome met.
Re-configure to a second layout.	Bill and Ana	Wednesday (week 2)	Increase available floor space/improve overall flow through work areas.	Expected outcome met.
Observe and listen to feedback from staff.	Staff, Ana, and Bill	Wednesday (week 2)	Gain buy-in.	Expected outcome met.
Move Decon. supplies into closet. Remove unused supplies. Remove one rack from room.	Claudette, Ana, Bill	Wednesday–Thursday (week 2)	Reduce traffic in decon. Room.	Expected outcome met.
Implement the best layout.	Ana, Bill	Thursday (week 2)	Measure improvements.	Expected outcome met.

Effects Confirmation

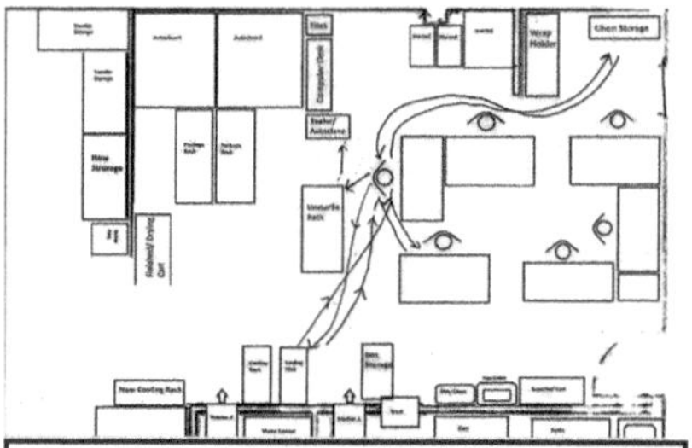

New layout decreased average travel distance to approximately 118 ft.
✔ 49.14% decrease in distance traveled.

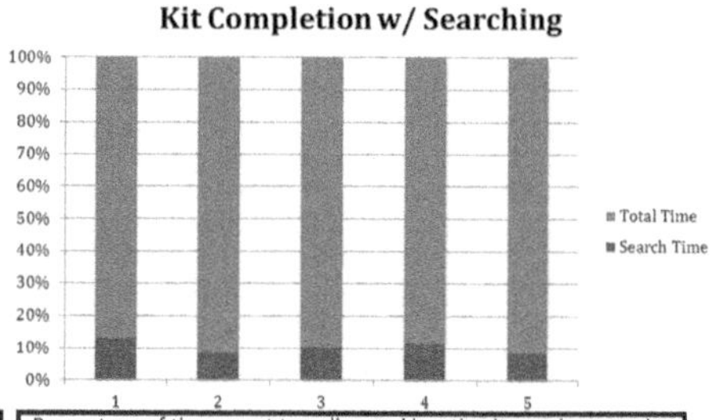

Percentage of time spent traveling and locating items decreased to approximately 10% of the total time.
✔ 61.54% decrease in travel and search time.

New Layout

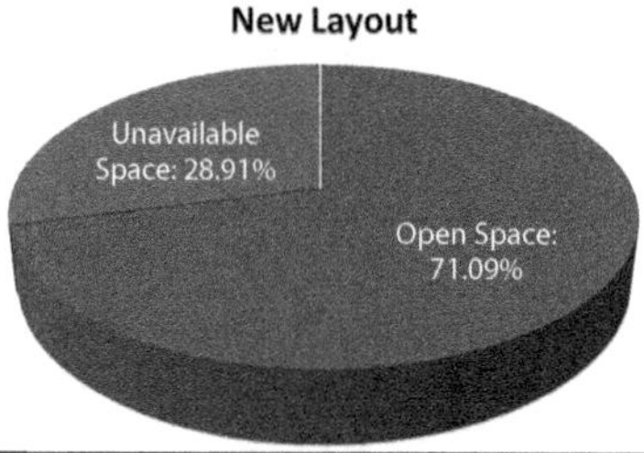

Open space available for travel and carts increased to 71% of the total floor space.
✔ 10.7% increase in available floor space.

General Observations:

- Points of contact with finished kits reduced by introducing the unsterile, ready to be loaded autoclave cart.
- Staff no longer has to cross each other's paths after separating workstations.
- A third cooling rack provides clearer task identification and improves the workflow.

Follow-Up Actions

Action	Responsible	To Complete Action	Due
Tape floor for unsterile area.	Engineering/ maintenance	Clean floor, remove old tape, and obtain new tape.	ASAP
Sign for visitors/deliveries on the second door.	Elsa	Create two signs.	ASAP
Move shelf unit in the center of the decontamination room to the wall.	Engineering	Remove the water system from the wall.	Friday (week 2)
Secure computer and phone cords to the wall.	Elsa and Eng.	Work order placed.	Friday (week 2)
Tape floor for cart storage.	Elsa and staff	Clean floor, remove old tape, and obtain new tape.	Friday (week 2)
Top shelf for linen rack.	Elsa	Order new 60″ × 24″ shelf and clips.	1 week
Move the marker board to the desk area wall.	Engineering	Work order placed.	1 week
Sort through the cart with miscellaneous items.	Elsa and staff	Sort cart and store unneeded items.	1 week
Organize storage shelves.	Elsa and staff	Acquire bins, dividers, and labels.	1 week

Case Study 6.3: Improving Workflow in the Medical Center ER	Date: XX/XX/20XX
Sponsor: ED	Authors: Grace and Ida

Background

- Two bay ER with curtain separation.
  - In 13% of emergencies, both beds are filled.
- The ER averages 50 visits per month over the past year.
- Nursing and provider staff are concerned with space available and both surplus and insufficient supplies in the ER.
- Traveling nurses, who are less familiar with the ER layout, are scheduled on average for 12 hours out of the 37 total nursing hours per day.
  - One 12-hour night shift.
  - 40% of ERs happen during the night nurse's shift.
- Often only one nurse from 10 p.m. to 6 a.m.

Current Condition

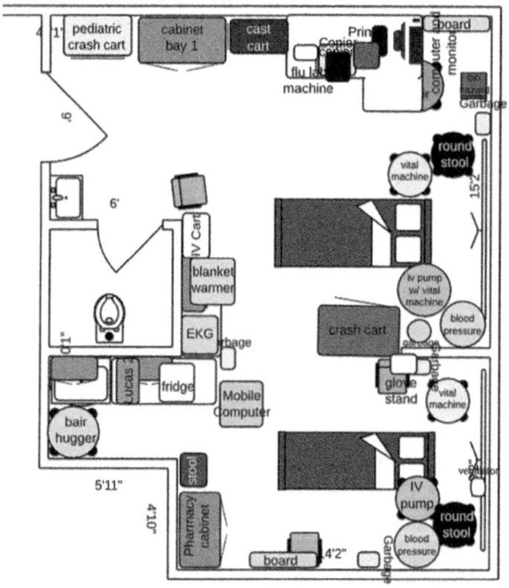

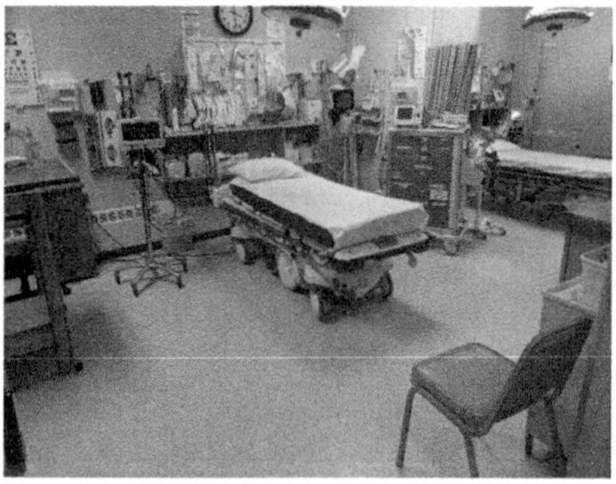

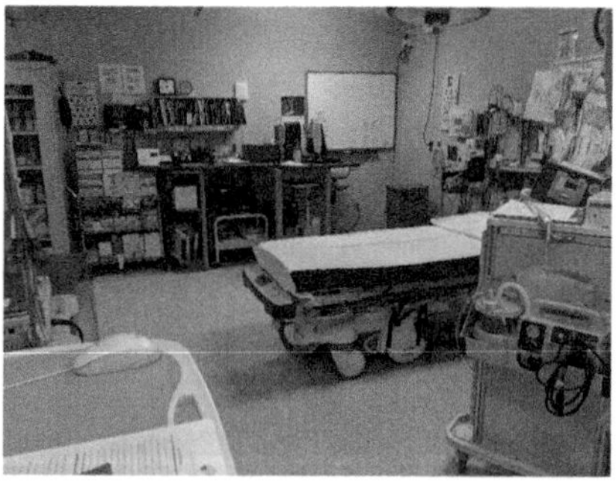

Workflow in ER: A 6-minute Mock Trauma ER Event of a compound fracture and patient going into shock with two registered nurses (RNs) and two certified nursing assistants (CNAs) was conducted to obtain a current state for the ER workflow.

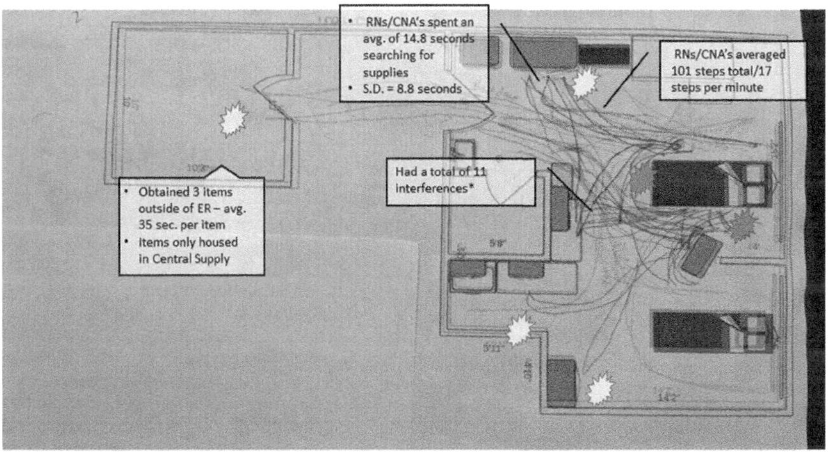

- The workflow prompts interference and search for supplies takes time from serving the patient.
  - 13% of the time two ER patients are admitted in the ER.
- Additionally, travel nurses aren't familiar with inventory placement in ER, making search time longer for needed items.

SMART Goals:	• Reduce average search time by 33%. • Reduce search time standard deviation by 50%. • Reduce interferences by 50%.

Root Cause Analysis

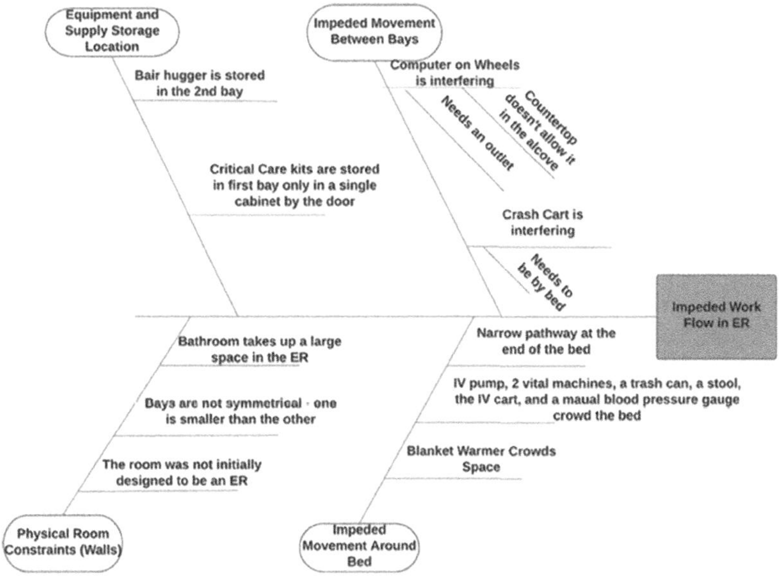

Proposed Countermeasures
Target Condition:

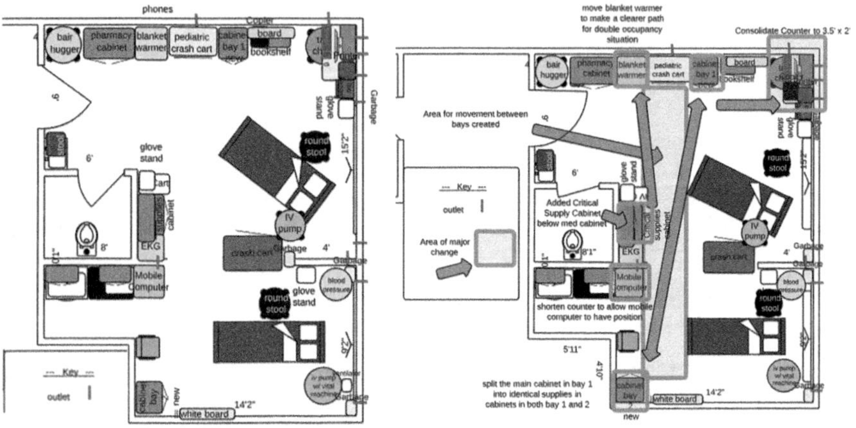

Implementation Plan

Action	Responsible	Deadline	Expected Outcome	Actual Outcome
Evaluate an improved ER layout with staff and management.	Grace and Ida; RNs, CNAs, providers on Mon. shift; Kate	Monday (week 2)	A layout is chosen to reconstruct the ER	Expected outcome met.
Create an inventory list with new supply levels and approve with staff.	Ida; Grace; RN and CNA staff on shift on Monday, Tuesday	Tuesday (week 2)	ER signage details supply locations (e.g., cabinet 1, bed 1, etc.)	Expected outcome met.
Adjust ER inventory based on the "ER Inventory" document.	Ida; Grace; nursing staff	Tuesday (week 2)	The ER Inventory is consistent with the levels approved by staff	Expected outcome met.
Construct the improved ER layout.	Ida; Grace; nursing staff; Kate; maintenance	Tuesday (week 2)	The improved layout is constructed in the ER with tape outlining the placement of equipment, shelving, cabinets, and counters	Expected outcome met.

Effects Confirmation

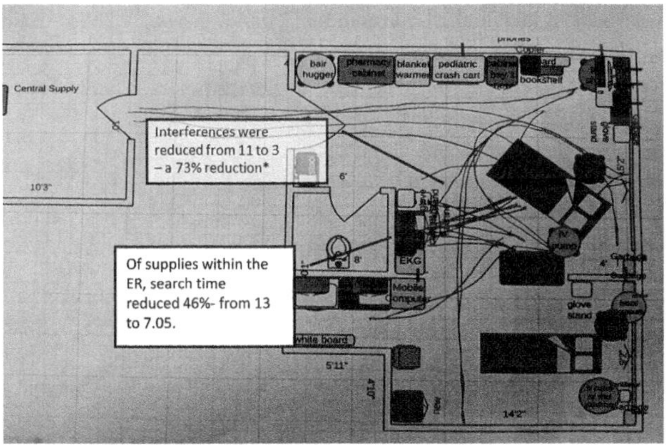

* *This trial had only two RNs instead of two RNs and two CNAs due to short staffing*

- Of supplies within the ER, **search time reduced 45.77%** – 13–7.05 seconds avg. time per item.
- **St.Dev. 43.48% reduction** – 7.52–4.25 seconds (not the 50% goal but close enough!).
- **73% reduction in interferences**.
- Staff were happy with the new layout.

Follow-Up Actions

Action	Responsible Party	To Complete Action	Due
• **Evaluate the new layout, and if approved, implement permanent cabinets/shelving.** • **Recommend replacing taller cabinets with two identical 2′ wide × 56″ tall × 18′ deep, identically colored cabinets for each bay, as well as one 33″ wide × 42″ high × 20″ deep cabinet for the critical supplies with an easily identifiable color, like red.**	Shirley and staff	Follow action.	In 1 week.

(Continued)

(*Continued*)

Action	Responsible Party	To Complete Action	Due
• **Evaluate the orientation of the workstation.** • **If approved, implement a permanent mirror at the workstation and/or swivel monitor.**	Shirley and staff	Follow action.	In 1 week.
• **Run scavenger hunts with all nursing staff and providers to familiarize the staff with the placement of supplies and equipment.**	Nursing staff	Follow action.	Next 2 weeks.
• **Install a counter over blanket warmer.**	Shirley and staff	Follow action.	In 2 weeks.
• **Install a cabinet/counter in the bathroom for urine and stool collection supplies.**	Shirley and staff	Follow action.	In 2 weeks.
• **Providers discuss and determine as a team which sutures need to be in the ER. For those determined necessary, create a storage system (*e.g., a portion of the existing shelving, a hardware style container, or another method*) and determine the placement of the sutures. For others, move to central supply.**	Providers	Follow action.	Next 3 weeks.

Case Study 6.4: Increasing Medication and Patient Scanning Compliance Rates	Date: XX/XX/20XX
Sponsor: Medical Floor	**Author: Arthur**

Background

- Eight Bed Critical Access Hospital.
- Serving a county of ~7,000 individuals, as per 2010 Census.
- Employs 20 Registered Nurses (RNs), 1 licensed practical nurse (LPN), and 9 certified nursing assistants (CNAs).
- A scanner and computer system available in each room for patient and medicine identification and charting that goes unused because:
  - National Drug Code's (NDC) barcodes aren't accessible, legible, or present on medication or patient.
  - The scanner is not feasible to grab in any room.
  - The patient becomes a barrier to the use of the scanner and computer system.
- Non-use can lead to patient safety, satisfaction issues:
  - Inaccurate Charting.
  - Untimely Charting.

Current Condition

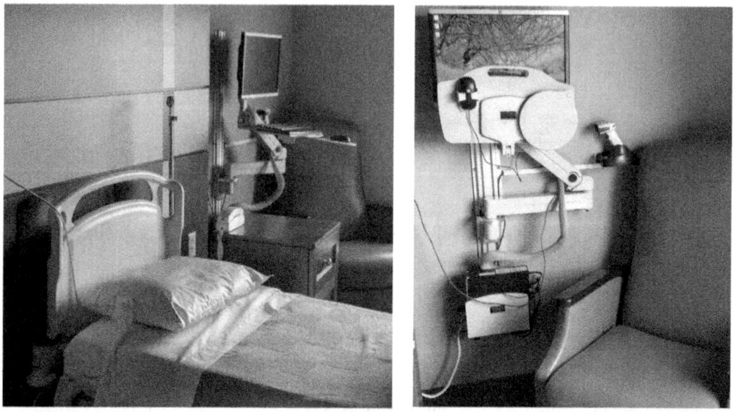

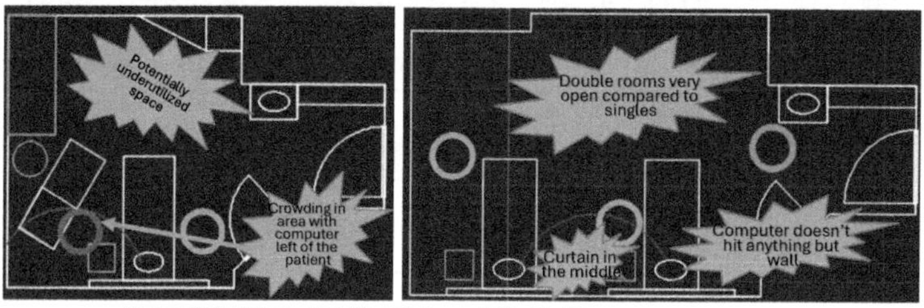

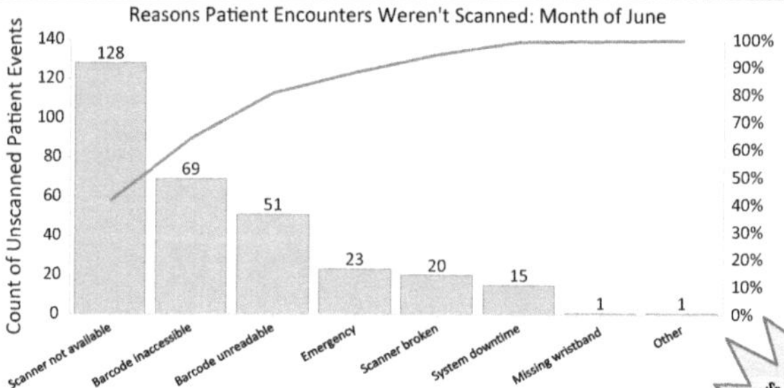

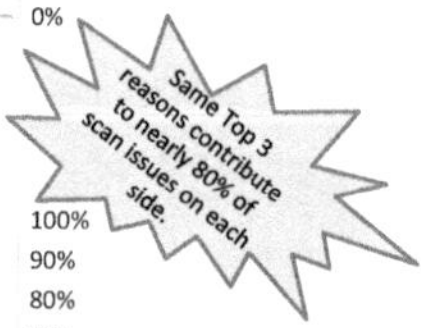

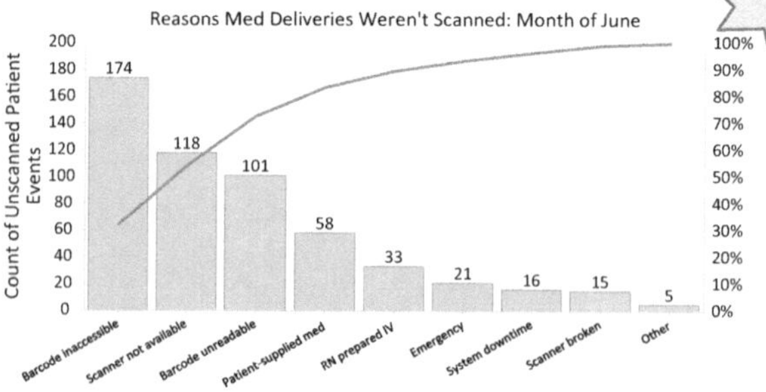

For June 20XX, there were 2051 Patient Interactions requiring some form of scan.

SMART Goals:	• Improve the patient scan rate by 10%. • Improve the med scan rate by 15%.

Root Cause Analysis

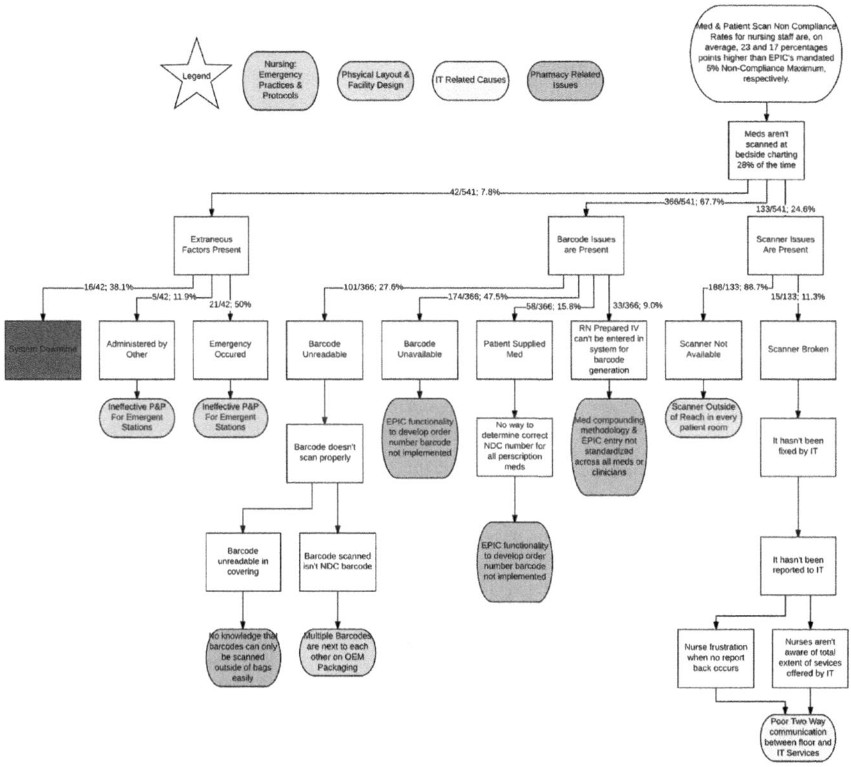

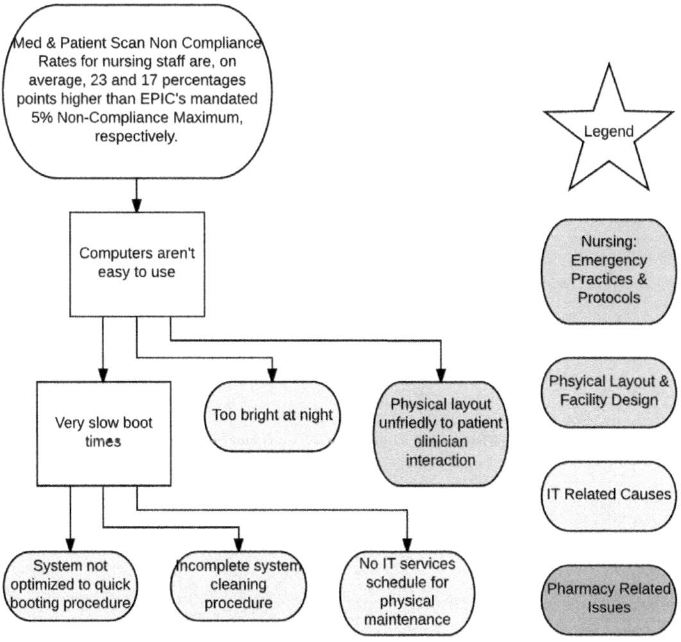

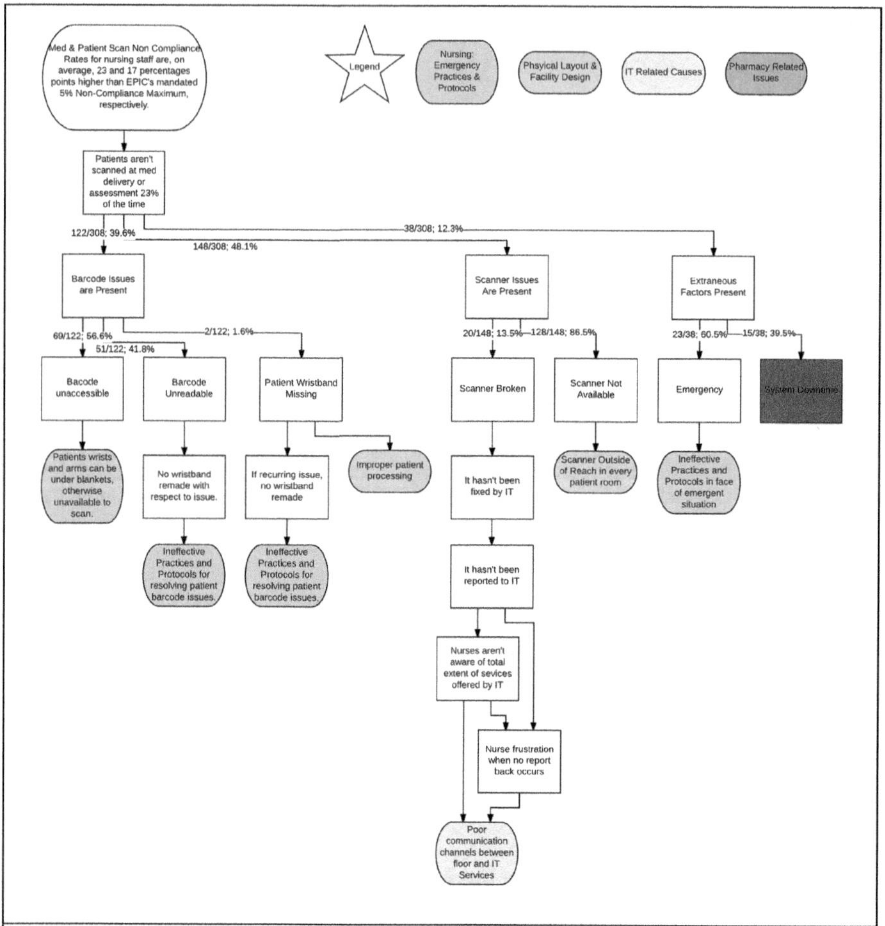

Proposed Countermeasures

IT Services Counter Measures

Root Cause	Potential Counter Measure	Responsible Party	Problem Area	Expected Results
Computer monitors too bright at night.	Pilot an install of F.Lux on patient room computers for testing.	Bertha, Arthur	Computer Systems	Usability of computers at night increases due to less eye strain, patient sleep interruption through auto-dimming and orange display at night based on latitude.

(Continued)

(Continued)

Root Cause	Potential Counter Measure	Responsible Party	Problem Area	Expected Results
System not optimized to quick booting procedure.	Diagnose reason.	Bertha, Arthur	Computer Systems	Reason determined to be due to Fast Startup toggle being ON, changed to OFF.
No IT services schedule for physical maintenance.	Develop a schedule for dust removal.	Bertha	Computer Systems	Physically Clean computers on a set periodic schedule to remove dust buildup on vents, heat sinks, fans, etc.
Incomplete system cleaning procedure.	Develop a procedure document for system cleaning, with a schedule for cleaning and document update.	Bertha, Cristobal, Arthur	Computer Systems	Install CCleaner, run registry cleaner and regular cleaner, and uninstall CCleaner at set frequency. If software changes, reflect such in procedure document.
Incomplete nursing–IT communication channels – INTERDEPART-MENTAL: Nursing.	Develop communication agreements between nursing and IT.	Arthur, Bertha, Dolly	Med Scan	Sign off on communication plan, methods, and expectations for each side of the coin between IT and nursing.

Pharmacy Departmental Counter Measures

Root Cause	Potential Counter Measure	Responsible Party	Problem Area	Expected Results
INTERDEPARTMENTAL: Nursing – multiple barcodes on some meds lead to incorrect scanning.	Education and communication about all meds currently in pharmacy, as well as new ones that come in, that have more than just NDC barcode.	Dolly, Fay	Med Scan	List of meds used that have two or more barcodes to differentiate between, as well as trouble shooting to find the correct barcode.
INTERDEPARTMENTAL: Nursing – incffcctivc communication scheme for med barcode scanning resolution.	Develop communication agreements between nursing and pharmacy.	Arthur, Fay, Dolly	Med Scan	Sign off on communication plan, methods, and expectations for each side of the coin between pharmacy and nursing.

(Continued)

Root Cause	Potential Counter Measure	Responsible Party	Problem Area	Expected Results
INTERDEPARTMENTAL: Nursing – no best practices for problematic med packaging scanning.	Communication of common med package types that are problematic to scan, and best practices on how to scan these meds.	Arthur, Fay, Dolly	Med Scan	List of packaging types, and best ways to resolve scan issues, such as scanning the bag lying down.
Epic functionality for order barcode for patient own meds not implemented at DLMed.	Contact Bertha, talk through the process.	Arthur, Gonzalo, Fay	Med Scan	Confirmation of the process for how to develop barcode labels for patient own meds; found out can be used for RN Prepared IV once Pyxis interface developed.

Nursing Department Counter Measures

Root Cause	Potential Counter Measure	Responsible Party	Problem Area	Expected Results
INTER-DEPARTMENTAL: IT – incomplete nursing–IT communication channels.	Develop communication agreements between nursing and IT.	Arthur, Bertha, Dolly	Med scan	
INTER-DEPARTMENTAL: Pharmacy – multiple barcodes on some meds lead to incorrect scanning.	Education and communication about all meds currently in pharmacy, as well as new ones that come in, that have more than just NDC barcode.	Dolly, Fay	Med scan	List of meds used that have two or more barcodes to differentiate between, as well as trouble shooting to find the correct barcode.
INTER-DEPARTMENTAL: Nursing – ineffective communication scheme for med barcode scanning resolution.	Develop communication agreements between nursing and pharmacy.	Arthur, Fay, Dolly	Med scan	Sign off on communication plan, methods, and expectations for each side of the coin between pharmacy and nursing.

(Continued)

(*Continued*)

Root Cause	Potential Counter Measure	Responsible Party	Problem Area	Expected Results
INTER-DEPARTMENTAL: Nursing – no best practices for problematic med packaging scanning.	Communication of common med package types that are problematic to scan, and best practices on how to scan these meds.	Arthur, Fay, Dolly	Med scan	List of packaging types, and best ways to resolve scan issues, such as scanning the bag lying down.
Ineffective practices and protocols for emergency situations.	Refine and post expectations of nurses scanning conduct in emergent situations.	Dolly, Gonzalo, Nursing Staff	Med scan Patient scan	Clear, visual, and easy-to-follow set of guidelines for how to operate under pressure of emergency situations.
Improper patient processing leading to no barcode wrist band.	Practice and protocol specifically outlined for barcode development for patient wrist band.	Dolly, Gonzalo, Nursing Staff	Patient scan	Outline of how and when the patient barcode should be developed and printed out, statement of openness to improvement within constraints of hospital, and consequences of continued non-compliance.
Incomplete patient barcode issue resolution guidelines.	Develop a list of patient barcode issue troubleshooting methods.	Dolly, Gonzalo, Nursing Staff	Patient scan	Document of common issues, ways to resolve most easily, timeliness expectations.

Target Condition:

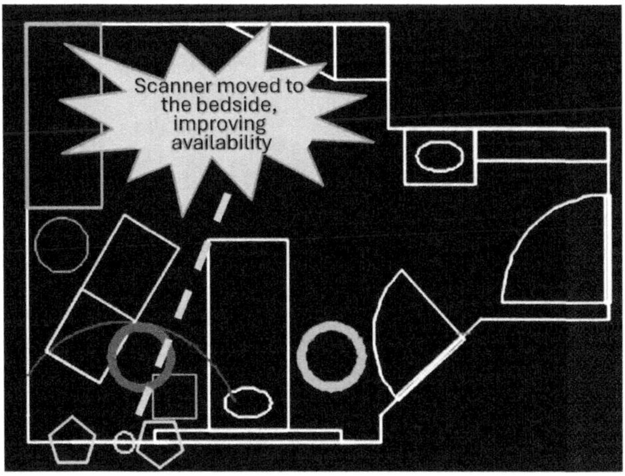

Implementation Plan

Action	Responsible	Deadline	Expected Outcome	Actual Outcome
Diagnose boot time issue.	Bertha, Arthur	Monday (week 2)	Boot time from power button press to accessible environment is reduced.	Boot time reduction by 66%.
Implement F.Lux for each patient room computers	Bertha, Arthur	Monday (week 2)	Automated dimming and screen coloration based on time of day.	Implemented.
Get opinion of nurses, nursing management, Physical Therapy (PT), Planetree on proposed room layout reorganization.	Arthur, Dolly, Gonzalo, Bertha (PT), Hanna	Monday (week 2)	Receive feedback on how to rearrange the room.	Went poorly, can't move furniture to ideal locations while maintaining functionality of patient information whiteboard, can't move white board anywhere else that will fit.
Determine meds not scanning correctly.	Arthur, Fay	Tuesday (week 2)	Identify all meds currently present in the formulary that scan incorrectly in terms of NDC.	Four high-use meds redefined in the software to scan correctly across DLMed; more to follow.
Formulate IT communication plan.	Bertha, Arthur, Dolly	Tuesday (week 2)	Development, reinforcement of the list of issues to contact Bertha, IT about in event of a problem with technical support systems.	Plan developed, awaiting sign off from Dolly.
Move all scanners to the left side of the computer monitor, and remount them.	Arthur, Isaias	Tuesday–Wednesday (week 2)	Remount all scanners for ease of access away from the patient recliner, closer to the computer body or frame.	On Tuesday: The patient moved into the pilot room; will move forward on the second with nurse and mgmt. approval. Wednesday Update: Decision made to move forward with plan, work order placed to remount scanners, patch holes in all single bed rooms.

(Continued)

(Continued)

Action	Responsible	Deadline	Expected Outcome	Actual Outcome
Gain more information from Hanna, Bertha, on order number printing of labels for non-charge patient meds.	Arthur, Dolly, Josephine, CFO	Wednesday (week 2)	Determine if the printer that's been setup for label printing is a viable solution for the patient billing process while remedying the patient meds issue.	Determined that the printer will be able to print labels for both patient meds as well as RN prepared IVs after the Pyxis is interfaced correctly.

Effects Confirmation

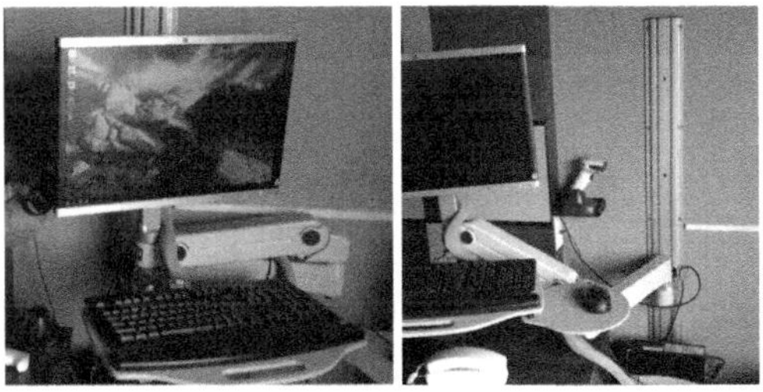

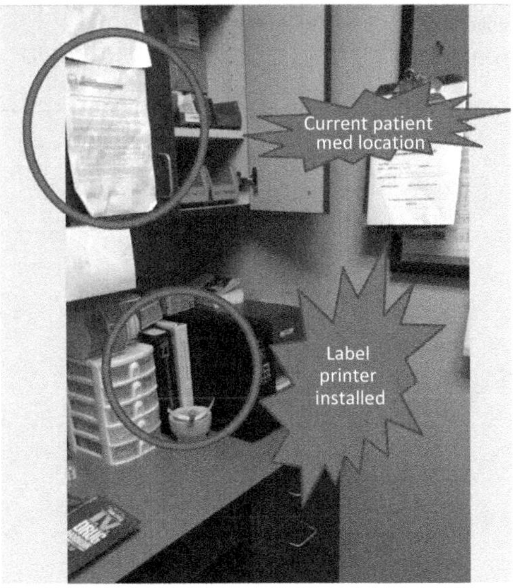

Major Accomplishments:

- Boot time reduction by 66%.
- Automated dimming and screen coloration based on time of day.
- Four high-use meds redefined in the software.
- Moved all scanners to left side of the computer monitor, and remounted them.
  - One day test in one room showed all Patient Encounters and Meds were scanned.
  - *Results may be biased because the individual's behavior could have changed due to being observed.*

Follow-Up Actions

Action	Responsible Party	To Complete Action	Due
IT communications plan sign off.	Bertha	Follow action.	Next Monday.
Pharmacy communications plan sign off.	Dolly, Fay	Follow action.	Next Wednesday.
IT maintenance schedule signed off.	Bertha	Follow action.	Next Friday.
IT review schedule developed.	Bertha	Follow action.	Next Friday.
Implement printer system for patient barcodes in other rooms.	Bertha, Dolly	Follow action.	In 2 weeks.
IT list of services document.	Bertha	Follow action.	In 2 weeks.
Deployment of common med package types that are problematic to scan, and best practices on how to scan these meds.	Dolly, Fay	Follow action.	In 2 weeks.
Formalize document showing the emergency workflow, with all necessary steps.	Dolly, Fay	Follow action.	In 2 weeks.
Formalize document listing out the patient scanning process.	Dolly, Fay	Follow action.	In 2 weeks.
Determine med and patient scan rates for portion of month under new systems.	Dolly, Fay	Follow action.	In 1 month.
Implement the RN prepared IV protocol once the Pyxis interface is developed.	Dolly, Fay	Follow action.	In 2 months.

Case Study 6.5: Reducing Missing, Misplaced, and Hard to Find I.V. Medications	Date: XX/XX/20XX
Sponsor: Pharmacy	**Author: Vicky**

Background

- The pharmacy technicians are responsible for stocking the Omnicell with the I.V.s needed for the day.
- These I.V.s include mixed and non-mixed I.V.s.
- The pharmacy technicians stock the Omnicells once a day with a batch run and also as needed throughout the day.

Current Condition

Problems

- I.V.s cannot be administered until nurses possess the I.V.
- If the nurse cannot find the medication they must waste time looking for the I.V.
- If no I.V. is found the nurse must go to pharmacy to find I.V.
- These actions cause wasted time for nurses, pharmacy, and pharmacy technicians.

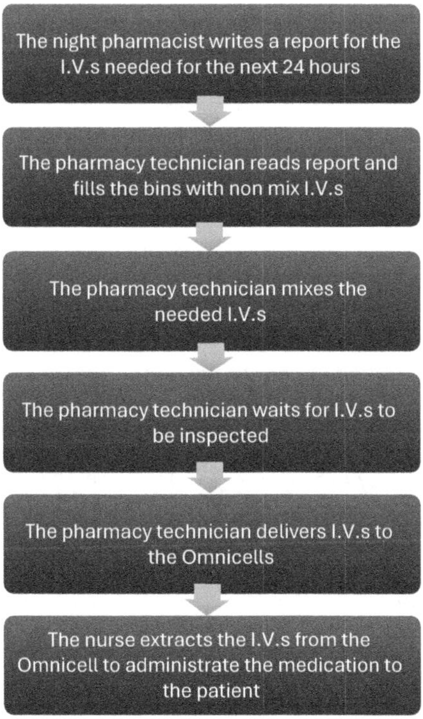

SMART Goal:	• Reduce the number of occurrences of I.V. medication "search and rescue" by nurses on the Medical Floor.

Root Cause Analysis

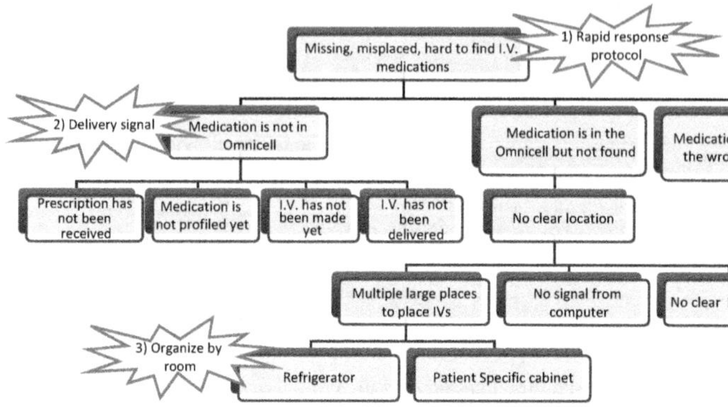

Proposed Countermeasures

1. Rapid Response Protocol
   - Document that states what to do when the I.V. medications are not found.
   - List of steps for the nurse's course of action to find the I.V., trigger root cause problem-solving.
2. Delivery Signal
   - Board that shows the last time the pharmacy technician was at the Omnicell.
   - Could also show that a medication has been delivered for a certain room.
3. Reorganization of I.V.s and the Patient Specific Areas
   - Create a specific location for each patient room in the refrigerator and patient specific area within Omnicell.
   - Label each location with the patient room number.
   - Pharm tech places I.V. meds in the specified location for each room number.

Countermeasures for the Proposed Organization

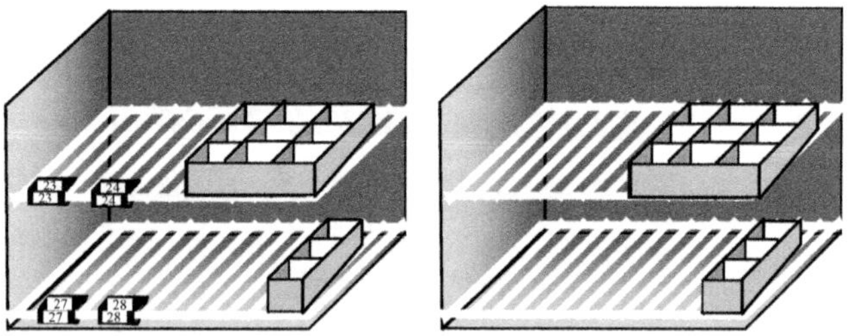

- Organize shelves with tape lanes and binder clips labeled with patient room numbers.
- Tape lanes used to designate sections in the refrigerator.
- Binder clips used to specify what rooms are in the section.
- Pharmacy technicians place I.V.s in the refrigerator in sequential order of patient room number, left to right, top to bottom.
- Nurses preserve the fridge organization when they retrieve I.V.s.

Pros:

- Saves search time for nurses and reduces chance of being unable to locate medication.
- Saves time for Pharmacy by reducing interruptions of nurses unable to find medication, and when Pharmacy must locate a specific I.V.
- Organized and flexible.

Implementation Plan

Action	Responsible	Deadline	Expected Outcome	Actual Outcome
Design an organization plan.	Vicky	Wednesday (week 2)	-	Binder clips.
Make the clips for room numbers.	Vicky	Wednesday (week 2)	-	Made 39 clips.
Discuss with nurses and pharmacy technicians.	Vicky	Wednesday (week 2)	-	Willing to try.
Discuss with Amy and Fay.	Vicky	Wednesday (week 2)	-	Willing to try.
Tape the lanes on the refrigerator racks.	Vicky	Wednesday (week 2)	-	Made six lanes in each refrigerator.
Train pharmacy technicians with a procedure plan.	Vicky	Wednesday (week 2)	-	Ready for pilot run.
Inform nurses of the plan.	Vicky and Fay	Wednesday (week 2)	-	Communicating with nurses and email for all nurses from Fay.
Pilot run.	Pharmacy Tech	Thursday–Friday (week 2)	-	Collect information on the results of the implementation.
Take a survey.	Vicky	Friday (week 2)	-	Ask nurses and pharmacy technicians about their experiences.

Effects Confirmation

To confirm the results, a survey was taken among nurses and pharmacy technicians.

Nurse Questions	Day Shift 7 Nurses	Night Shift 2 Nurses
1. Were you able to find the I.V.s more quickly after organization?	Yes	Did not notice
2. Do you prefer the new organization to the original?	Yes	Did not notice
3. Were all the I.V.s organized according to the plan?	Yes	Did not notice
4. Did the refrigerator stay organized?	Yes	No

Pharmacy Technician Questions	Clarissa	Sarah
1. Does the room number labeling increase workload substantially?	No	No
2. Do you prefer the new organization to the original?	No effect	Yes
3. Do you have any other improvement ideas?	Bigger refrigerators	Dry eraser instead of binder clips
4. Did the refrigerator stay organized?	Day shift but not the night shift	

Follow-Up Actions

Action	Responsible Party	To Complete Action	Due
Ask the charge nurse to inspect to make sure the refrigerator stays organized according to the plan; follow up with nurses as needed.	Fay	Follow action.	Ongoing.
Continue the pilot study for one more week; make adjustments as needed using technician and nurse input.	Fay and Pharmacy Tech	Follow action.	In 1 week.
Communicate with all nurses (esp. night shift) about maintaining I.V. organization in the refrigerator.	Fay	Follow action.	In 1 week.
Implement a procedure for tracking missing medications. This will allow the pharmacy to identify the source of errors more easily, and work on eliminating other potential sources of error.	Pharmacy Tech	Follow action.	In 2 weeks.
Develop a standard "rapid response protocol" that nurses should follow in the case that they are unable to locate their patient's medication.	Fay and Pharmacy Tech	Follow action.	In 2 weeks.
Consider developing a non-intrusive way to notify that medication has been delivered to the Omnicell.	Fay and Pharmacy Tech	Follow action.	In 1 month.

7 Charting and Charge Capture Case Studies

Nurses, providers, technicians, and other healthcare professionals are primarily concerned with patient care. And yet, the business side of a hospital is crucially important for recovering supplies, expendables, pharmacologic, and personnel costs. The challenge, of course, is that the tasks necessary for being able to charge for supplies used and services rendered can be viewed as secondary, and the systems and workflows implemented are not very robust. The RIE briefs in this chapter focus on these back-of-house processes that are nonetheless critical for hospitals to stay in business. The case studies fall into two categories. The first two cases concern complete and accurate charting to allow timely, complete, and accurate bill coding. The last three cases deal primarily with capturing the medical supplies used during patient visits so that the hospital can bill for them.

Case 7.1 involved converting to bedside charting in the hospital's emergency department (ED). The changes implemented cut the number of charge-related charting errors in half while achieving a 40% reduction in charting time compared to the old practice of waiting for a break in the action or end of shift to chart. Case 7.2 deals with improving the accuracy of patient registration and medical charting so that the right information is captured in the electronic health record.

The new processes dramatically reduce the amount of time the billing office spent chasing down missing information or correcting errors and miscodes, resulting in a significant reduction in excessive accounts receivable days.

The focus of Case 7.3 is an ED where charges for most medical supplies were not captured at all. By taking advantage of the online charge capture feature of their current electronic health record system, developing procedures for capturing charges, and training staff, the department went from capturing less than 5% of medical supplies used to billing for more than 95% of them.

Cases 7.4 and 7.5 also involve medical supplies charge capture in EDs but are interesting contrasting cases. In one, the current system involved removing barcode stickers from supplies and sticking them to sticker sheet for later entry while the countermeasure was to replace the sticker system with a charge sheet. In the other, a charge sheet was replaced by a sticker system! But interestingly, both cases realized significant improvements in charge capture, which illustrates how the best countermeasure to address an issue is situationally dependent and how important it is to devise countermeasures that directly address the underlying causes in that situation rather than jumping to a favorite solution.

A common theme in these case studies is how many errors can be reduced when charting and documentation occur immediately, or at least very close in time, to when the service is rendered or the supply is used. Delaying these types of tasks opens the door for errors and omissions, so providers' work should be designed to facilitate timely charting.

DOI: 10.1201/9781003670698-7

Case Study 7.1: Documentation Process for the ED	Date: XX/XX/20XX
Sponsor: ED	Author: Nicole

Background

- If documentation is not done properly, a patient cannot be charged for the medication or procedure.
- Health Information Management (H.I.M.) is in charge of billing the patient.
- The charts the nurses have to fill out include two ED charts, a pharmacy and central supply sheet, an H.I.M. charge sheet, and a follow-up care sheet.

Current Condition

- Five (5) out of the 54 charts (9.26%) over 3 days from the ED had errors, which resulted in time traveling or excessive research to clarify the documentation.
- Such errors have occurred:
  - Vitals written on napkins.
  - Times for the nurse's charts and charge sheet do not correspond.
  - A patient's paperwork included in another patient's charts.

Charge capture

1. Physician diagnoses the patient's condition and writes the order on the chart.
   a. X-ray.
   b. CT.
   c. Lab.
   d. Medication.
   e. Wound dressing.
2. The nurse or nurse technician must order the test.
   a. **For X-ray, CT, and Lab**: The nurse places the order in the computer and calls the corresponding technician to order the test.
      i. The nurse or technician must initial and write the time the order was taken on the chart.
   b. **For Medication**: The nurse either contacts pharmacy or obtains medications from the Pyxis system.
      i. The nurse initials and writes the date/time the medication is administered on the chart.
      ii. For chargeable medication, the nurse takes the sticker and places it on the pharmacy sheet which is the second-to-last page of the charts.
   c. **For Wound Care**: The nurse signs and initials when the care was taken.
      i. For chargeable items used in wound care, the nurse places the sticker of the item in the central supply sheet, which is on the same page as the pharmacy sheet.
3. The nurse must get the patient's vitals before discharging the patient.
   a. If vitals are not documented on the chart, they are written on the hand or in a separate document, which will be added to the chart before the documentation is complete.
4. After all care toward the patient is completed, the nurse fills out the charge sheet.
   a. If ED is busy, this step is put off until the department calms down.
      i. This can be hours after the patient is checked out.
   b. If the patient's vitals were not already documented, they are documented now.
      i. Have to track down where the vitals were written down.
   c. The nurse initials all the care that can be charged toward the patient on the charge sheet.
   d. The nurse signs the paper to validate the charge sheet.

5. H.I.M. takes the chart to code the visit.
   a. Depending on where the nurse initialed determines the level of care the patient received.
   b. Must make sure all the items on the nurse's sheet and charge sheet are the same.
      i. What tests ED ordered.
      ii. Time the medication was delivered.
   c. If the sheets do not match.
      i. Go to ED to see if this can be cleared up; but by this time, usually the nurse and physician are not the same.
      ii. Cannot charge patient for care/medication/tests given.

SMART Goals:	• Reduce the amount of errors from the ED from 9.26% to less than 5% (46% reduction). • Reduce the amount of time H.I.M. personnel spend clarifying paperwork (avg. 4 minutes) by 25%.

Root Cause Analysis

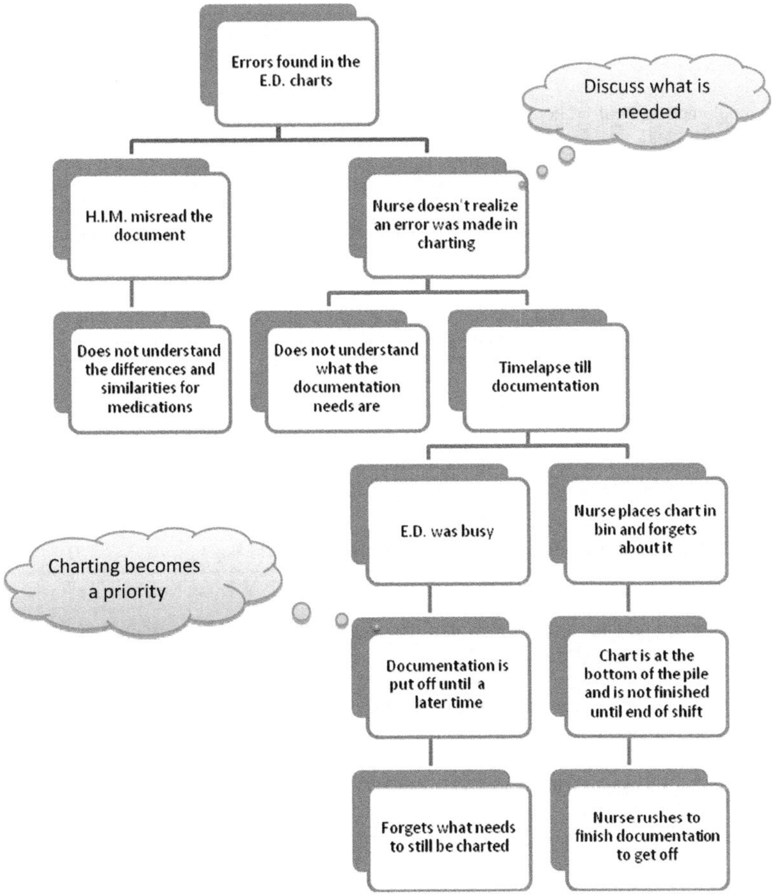

Proposed Countermeasures

1. Discuss what is needed between departments.
   - Discuss with nurses what H.I.M. needs for the charting to be valid.
   - Charting needs to be done on the charts not on separate documents.
   - Visual work instructions on a properly documented charge sheet.
2. Real-time charting
   - Charting is done when care is given at the bedside.
   - Convert "nurse's charts to be finished" bin to "dictated charts."
   - Finish the charts before a new patient is administered and before placing in dictation pile (estimated completion time = 4 minutes).
   - After dictation, the physician places the chart in the "dictated charts" bin.
   - Ensures documents are done promptly.

Implementation Plan

Action	Responsible	Deadline	Expected Outcome	Actual Outcome
Practice new bed side charting.	Nicole and Staff	Monday (week 2)	All charts completed when the patient is discharged.	All charts were completed when the patient was discharged.
Pilot implementation of real-time charting.	Nicole and Staff	Tuesday–Wednesday (week 2)	Staff practiced real-time charting.	Staff practiced the real-time charting.
Discuss the new charting with staff.	Nicole and Staff	Tuesday–Wednesday (week 2)	Gather staff input.	Staff thinks the new procedure will help.
Discuss final finding.	Nicole and Owen	Thursday (week 2)	Gather staff input.	Meeting is arranged.

Effects Confirmation

Nurse Questions	Nurses ($n = 3$)
1. Did the new process affect your work in a negative way?	No
2. Do you prefer the new procedure?	Yes
3. How many charts did you have to finish at the end of the day?	0

Number of charge capture errors during the pilot: 2 errors out of 42 charts. **48.6 % reduction!**

- Average time to finish charts: 2 minutes 15 seconds. **43.75% reduction!**

Follow-Up Actions

Action	Responsible Party	To Complete Action	Due
Continue to collect data on the charge capture errors that occur: • Physician. • Nurse and nurse technician. • Error type.	Owen	Collect data on the charge capture errors that occur: • Physician. • Nurse and nurse technician. • Error type.	Daily for the next 30 days.
Encourage staff to remind each other to chart in real time.	Staff	Encourage staff to remind each other to chart in real time.	Daily for the next 30 days.
Continue real-time charting.	Staff	Continue real-time charting.	Daily
For the new charge sheets: • Review proposed sheets with H.I.M. and nursing before finalizing. Specifically discuss the impact on charting and charge capture workflow. • Create visual work instructions for use in training. • Educate all impacted staff prior to implementation. • Lend support to H.I.M. and nursing during the transition to the new charge sheets.	Owen and Staff	• Review proposed sheets with H.I.M. and nursing before finalizing. Specifically discuss the impact on charting and charge capture workflow. • Create visual work instructions for use in training. • Educate all impacted staff prior to implementation. • Lend support to H.I.M. and nursing during the transition to the new charge sheets.	In 1–2 weeks
Explicit follow-up by Owen to assess whether real-time charting continues. If it does not, pursue root cause to discover why and conduct PDSA.	Owen		In 2 months.

Case Study 7.2: Business Office Accounts Receivable	Date: XX/XX/20XX
Sponsor: Business Office/ED	**Author: Bret**

Background

- Medical Center (MC) is a small critical access hospital that serves the town as well as the surrounding rural communities.
- A large majority of the MC's nighttime ED staff are non-permanent traveling staff.
- The MC serves 3–4 patients per day on average in the ED.
- The MC has recently encountered issues with documentation procedures for patient check-in, which have resulted in faults in the billing procedures, excessive time between treatment and billing, and an increase in dissatisfied patients due to improperly processed bills.

Current Condition

- *Data was collected during normal hospital operating conditions for a week (week 1).*
- *Daily data was collected to include all patient encounters in the ED that occurred within each 24 hours work day (7 a.m.–7 a.m.).*

Current State Value Stream Map (without Timeline)

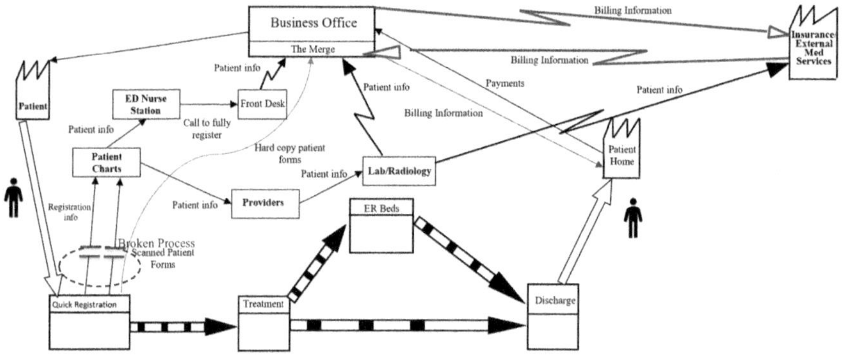

Distribution of Documentation Errors

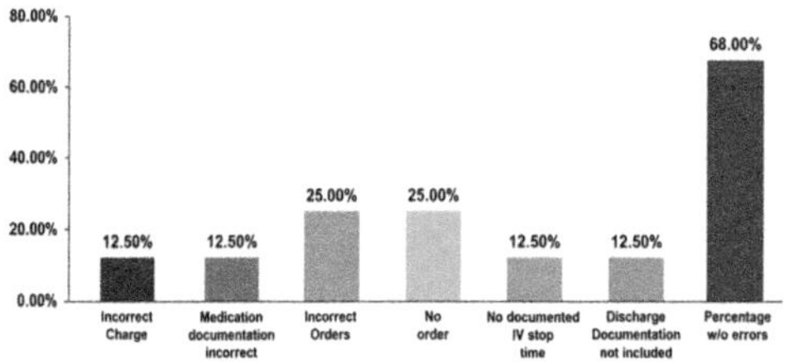

Key Stats:

→ 32% of ER visits had documentation errors.
→ Total of 24 patients seen in ED over collection period.
→ 8/9ths of all errors occurred in the ED.
→ None of the documentation-based billing issues were resolved.
→ $2907.50 (for 1 week) could not be billed out due to errors.

Major Problems:

- Lack of standard procedures across all ED nursing staff.
- Interdepartmental communication gaps slow fixing of documentation issues.
- Patient documentation scanning and naming methods are not consistent.
- Full registrations are not completed on a consistent basis in ED.
- Business office staff cannot bill efficiently and on time due to errors in documentation.

SMART Goals:	• Reduce ED visits with documentation errors to 15% or less. • Increase the number of issues resolved in two weeks or less from 0% to 25% of issues.

Root Cause Analysis

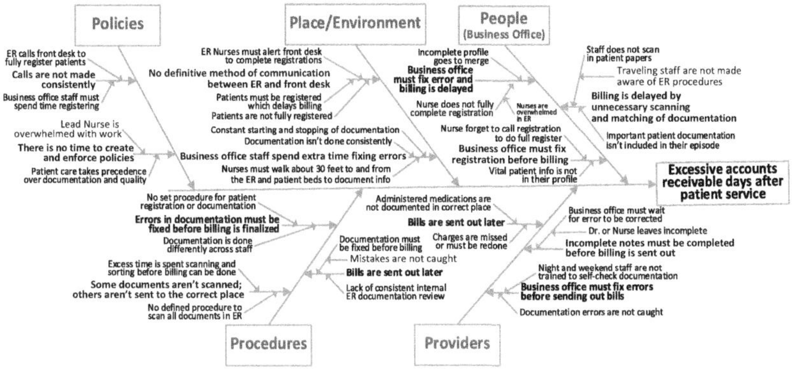

5 Whys:

Problem Statement 1: Excessive accounts receivable days after patient service.

- **Why?** Business office is unable to submit bills on time.
    - **Why?** Registrations from ED are incomplete or incorrect.
        - **Why?** Registrations and documentation were not reviewed after the shift ended.
            - **Why?** Given high workload, staff were unable to delegate time.
                - **Why?** The department must bring in traveling staff.
                    - **Why?** The department is understaffed.

Problem Statement 2: Excessive accounts receivable days after patient service.

- **Why?** The business office is unable to submit bills on time.
    - **Why?** Had to modify and change charges on the patient's chart.
        - **Why?** Charges were not dropped from their chart before discharge.
            - **Why?** Staff were not fully aware of system function.
                - **Why?** There is no training for a standard set of procedures and expectations.

Problem Statement 3: Excessive accounts receivable days after patient service.

- **Why?** The business office is unable to submit bills on time.
    - **Why?** Must sort through patient files and add documentation.
        - **Why?** Documentation was not scanned in or was scanned to the wrong place.
            - **Why?** ED staff was not aware of proper procedures.
                - **Why?** Staffing is not trained on procedures.
                    - **Why?** Lead staff does not have time to complete training.

Problem Statement 4: Excessive accounts receivable days after patient service.

- **Why?** The business office is unable to submit bills on time.
    - **Why?** Must fully register previous patients from the ED.
        - **Why?** Patients were not fully registered at the time of admittance.
            - **Why?** Front desk was not made aware of their registration.
                - **Why?** ED did not communicate with registration.
                    - **Why?** Procedures for communication are not effective.

Proposed Countermeasures

Countermeasure	Steps	Responsible	Goal
1. Registration and documentation starter flow sheet.	• Discuss proper info with nurses. • Make a flow sheet and check with staff. • Laminate and place at each nurse station, have senior staff discuss and support the new sheet.	ED nurses, Bret ED nurses, Bret ED nurses, Bret	Set in place a set procedure for all travelers and staff to follow for documentation during a shift.
2. Scanning instruction and formatting sheet.	• Discuss proper procedures with ED staff. • Create short guide/flow sheet with staff. • Revise to assure accuracy and have upper staff discuss and support new procedures to scan all documents, not use paper box. • Laminate and place at nurse stations.	ED nurses, Bret ED nurses, Bret ED nurses, Bret ED nurses, Bret	Orient staff to scan all essential documents into patients' charts and label them for easy reference by the business office. Get rid of paper stack that takes time to process during the merge.
3. Daily documentation audit by a staff member to check and correct errors.	• Discuss duties with full-time senior faculty and who will perform check. • Discuss time of check and how long it will take. • Have a once or twice daily check of documentation.	Arlene, Bret, Cindy, Don Arlene, Bret, Cindy, Don Arlene, Bret, Cindy, Don	Serve as a second line of defense to assure documentation is not missed and all charts are filled correctly, less issues for business office to fix.
4. Checklist for documentation that must be completed before a patient in the ED can be discharged.	• Get feedback from nurses and staff on processes and create form. • Implement with each nurse checking over documentation after each patient.	Full-time ED nurses, Bret Full-time ED nurses, Bret	Assures nurses check all documentation is correct before finishing with each patient. Serve as certification for quality and lead staff to monitor proper documentation practices.
5. Create a cross-departmental communication system with whiteboards or suggestion boxes.	• Order corkboards through materials. • Place boards in each department with comment forms and have Harvey check them each day.	Materials, Bret Materials, Bret	Increase communication between the departments and the CEO; encourages openness and problem-solving in the workplace.
6. Enforce new policy for all staff to contact the front desk via Teams and calls to fully register patients in the ED during business hours.	• Discuss policy with ED and reception. • Make changes as necessary. • Have all ED staff call the front desk to fully register people during business hours.	Full-time ED nurses, Bret Full-time ED nurses, Bret Full-time ED nurses, Bret	Have patients fully register before the merge so the business office doesn't need to spend time registering them.

Target Condition:

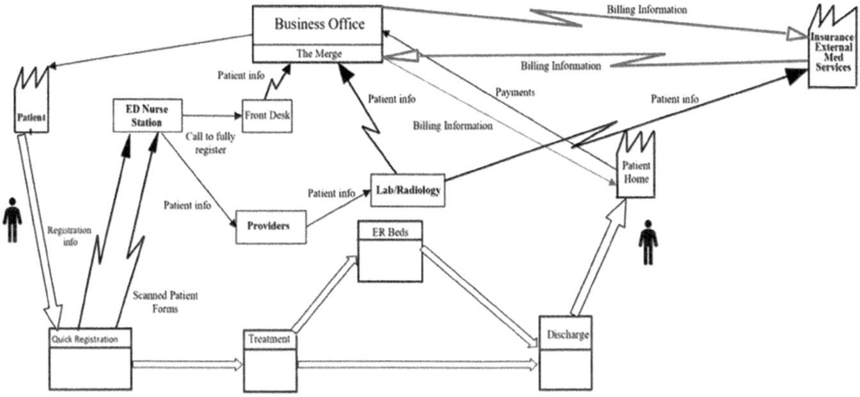

Implementation Plan

Action	Responsible	Deadline	Expected Outcome	Actual Outcome
Create drafts of guidance flow sheets for documentation and scanning and obtain nurse feedback.	ED staff, Don, Bret	Tuesday (week 2)	See Countermeasure #2.	Expected outcome met.
Discuss audit and who will perform it with staff; have Don emphasize the importance of that check.	Bret, Don, Nurses	Tuesday (week 2)	See Countermeasure #3.	Expected outcome met.
Finalize guidance sheets and print and place at nurse stations; brief staff on use.	ED staff, Don, Bret	Wednesday (week 2)	See Countermeasure #2.	Expected outcome met.
Have staff perform twice daily check through of documentation to check for errors.	Bret, Don, ED staff,	Wednesday (week 2)	See Countermeasure #3.	Expected outcome met.
Create a documentation checklist and obtain nurse feedback; print and post at nurse stations.	Bret, ED Staff, Don	Wednesday (week 2)	See Countermeasure #4.	Expected outcome met.
Have nursing staff begin calling to fully register the ED patients during business hours.	ED staff, Bret, Reception	Wednesday (week 2)	See Countermeasure #6.	Expected outcome met.

(*Continued*)

(*Continued*)

Action	Responsible	Deadline	Expected Outcome	Actual Outcome
Have nurses and ED staff perform checks using the checklist after each patient; check that checklists are being filled out after each patient. Checklists are collected by head nurse and quality head for review during merge.	Bret, Don, Franklin	Thursday (week 2)	See Countermeasure #4.	Expected outcome met.
Order cork boards through materials and place in each department.	Bret, Gert	Friday (week 2)	See Countermeasure #5.	Expected outcome met.
Post forms on feedback boards and have staff start using boards.	Bret, Harvey	Friday (week 2)	See Countermeasure #5.	Expected outcome met.

Effects Confirmation

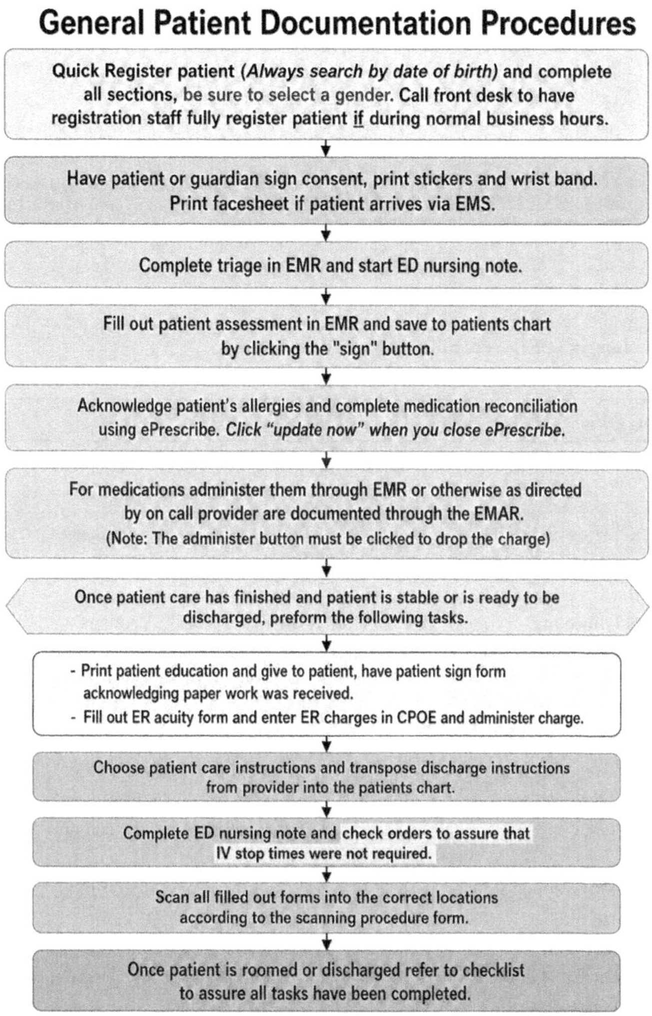

Changes Made During Implementation:

- The flow chart was edited to make the important actions stand out more to the nurses.
- The scanning info sheet was edited to include more accurate information using business office and nurse feedback.
- The checkoff sheet was modified to serve as way to document errors by upper management.
- Organized checkoff sheets in the central position at the nurse station for easy access.
- Had the head of quality do a daily check of nurses' documentation.

General Observations:

- Staff was very receptive to changes, open to giving them a try.
- Staff suggested ways to change documentation given potential future changes.
- Staff check off sheets regularly after patients.

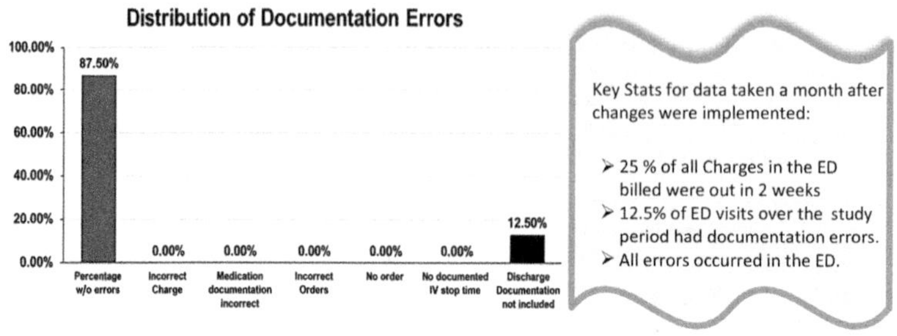

Goals Met:

- **25% of charges in ED were billed out in 2 weeks or less!**
- **Only 12.5% of ED visits over the study period had documentation errors.**

Follow-Up Actions

Action	Responsible Party	To Complete Action	Due
Post feedback boards in the ED, clinic, lab, registration, business office.	Harvey	Put boards up in departments with feedback sheets posted on boards.	In 1 week.
Schedule weekly meetings between the business office and ED/nursing staff.	Harvey, Don	Begin scheduling meetings between ED staff/nurses and the business office to discuss issues, how to solve them.	In 1 week.
Hire a staff member to perform daily audits on ED documentation to assure consistency as well as train nursing staff.	Harvey, Don	Hire a staff member and train them in the software to check and quality check the documentation at the end of each shift.	In 1 week.
20-minute refresher training for all ED staff new and old on documentation and full registration and what is needed/essential for the business office to function the best. Want to train all ED staff to fully register.	Irma, Jose	Schedule a time for a training with full-time ED staff to refresh on what is needed and essential for proper coordination with the patient.	In 1 week.
Make Pro fees into a digital paper/fillable pdf that can be used by all providers and be easily shared between business offices.	Katia, Jose, Dr. Philippe, Maria	Discuss creating an online version of the form with IT, clinic, and business office. Gain doctor approval and implement forms into system.	In 1 week.
Implement policy for all charges to be dropped and charged as soon as they come through after registration.	Katia, Lee, Maria, Nate	Work with one another and ED to get caught up on charges and errors and create a continuous flow of charges being dropped as patients get taken care of.	In 5 weeks.

Case Study 7.3: MC Charge Capture and Billing	Date: XX/XX/20XX
Sponsor: Radiology and ED	**Author: Andrea**

Background

- The MC is a medium-sized critical access hospital that mainly serves the surrounding communities to the north and south, as well as the local population and large tourism contingent.
- Currently, the administration staff is having issues with charge capture in the ED and radiology, and would like to ensure that all chargeable items that are being administered are being recorded for charges.
- The ED on average sees 15–20 visits per day.
- Fully online charge capture systems are in the midst of being implemented to eliminate the old system, but a new system is needed to ensure that online charges are consistently documented.
- Inventory is hand-counted quarterly to check the accuracy of the inventory management system and recorded 99% accuracy in the most recent quarter.

Current Condition

Assumptions:

- *Inventory functions with a Kanban style pull system, with empty boxes set aside and then scanned to restock inventory and determine charged items each morning.*
- *ED and Radiology have inconsistent patient volumes and thus chargeable item usage is random.*
- *Data collected during a normal week of operations.*

Current State Value Stream Map (without Timeline)

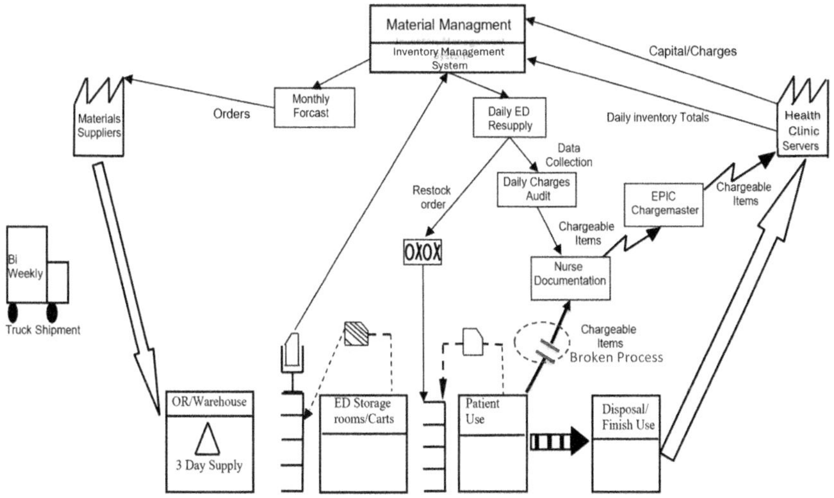

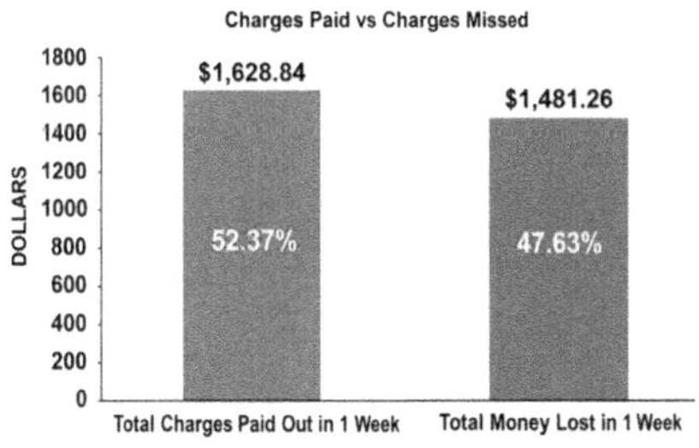

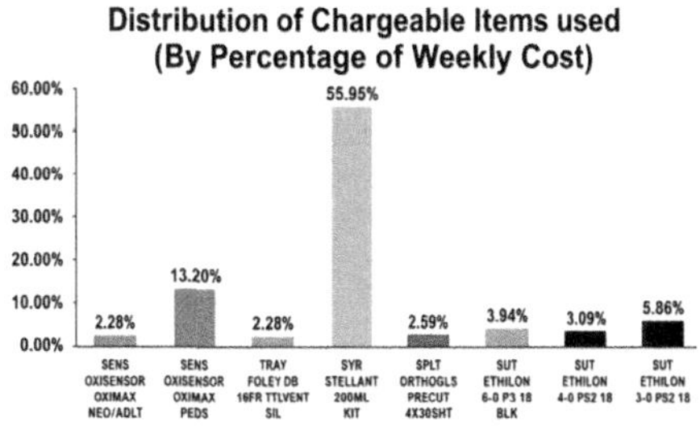

Key Stats:

- $3,096.24 worth of supplies used in a week.
- Estimated loss of $1,481.60/business week, which extrapolated out to $108,156.80/year.
- 96.14% of charges billed came from CT, while 3.86% came from ED.
- **3.13% of all ED materials used were charged.**
- On average $73.30/patient per day were lost over the study period.

Note: Charge amounts do not include non-material-based ED charges, so the actual number is likely higher.

Major Problems:

- The current charge capture system using physical stickers is outdated and prone to errors and inconsistencies.
- Nurses have not included charge capture a part of their normal documentation procedures.
- Lack of effective or consistent labeling and naming of chargeable items.
- No standard procedures for charge delivery across ED in software or general documentation.
- No area for triage prior to ER room entry, slows throughput, and causes ER nurses to do more work.

SMART Goals:	• **Increase the percentage of all chargeable materials charged by 20%.** • **Increase the percentage of chargeable ED materials charged by 20%.**

Root Cause Analysis

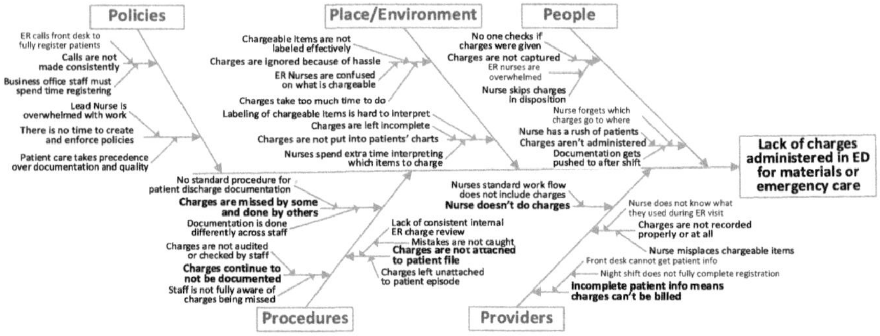

Problem Statement 2: Lack of charges administered in ED for materials or care.

↳ **Why?** Charges not stickered and recorded onto patients' sheet/documentation.

↳ **Why?** Nurses do not enter charges in a consistent manner.

↳ **Why?** Lack of standardized flow through Epic.

↳ **Why?** Charges are not well incorporated into discharge flow.

↳ **Why?** No training to set agreed upon standard ED patient procedures through Epic.

Problem Statement 3: Lack of charges administered in ED for materials or care.

↳ **Why?** Charges not stickered and recorded onto patients' sheet/documentation.

↳ **Why?** Nurse is unfamiliar with actual names of chargeable items.

↳ **Why?** Chargeable items were labeled based on inventory names.

↳ **Why?** Limited Epic access to changes prevented concise naming.

↳ **Why?** Nurses were not originally consulted on charge capture naming.

Problem Statement 4: Lack of charges administered in ED for materials or care.

↳**Why?** Charges not stickered and recorded onto patients' sheet/documentation.

↳**Why?** Nurses do not enter charges in a consistent manner.

↳**Why?** The system of charging adds extra time to documentation discharge.

↳**Why?** Charging stickers and online charges do not flow easily with current procedures.

↳**Why?** Nurses must document patients and then record and click charges for documentation.

↳**Why?** The current system does not integrate charges into charting in an effective manner.

Proposed Countermeasures

Countermeasure	Steps	Responsible Party	Goal
1. Reference sheet to clarify and define all Epic material charge labels for nurses.	Create sheet. • Adjust with nurses assistance. • Print, laminate, and place at all nurse stations.	Andrea, ED Nurses Andrea, ED Nurses Andrea, ED Nurses	Allow nurses to quickly and easily decipher which charges are for which materials.
2. Standardize the Epic format used by all nurses, including the location of charges in the navigator and order of operations.	• Discuss changes with nurses and head nurses. • Develop a standard format in current Epic. • Work with all nurses individually to change their Epic format; also place notice in the ED nurse station.	ED Nurses, Andrea, Barry, Erin ED Nurses, Andrea, Barry, Erin ED nurses, Andrea, Barry, Erin	Having all staff run disposition procedures the same way with charges in the same place. The goal is to make charges a habit and document consistently.
3. Implement online charge capture through Epic and eliminate the sticker system in ED.	• Discuss the start date of the new online system with materials and head nursing. • Have all nurses document charges.	Barry, Erin, ED Nurses, Andrea, Radiology Staff Barry, Erin, ED Nurses, Andrea, Radiology Staff	Allows for the nurse to administer all charges in a convenient manner right before disposition of the patient.
4. Create and run specific training on which items are most used and how to charge for all new and old employees.	• Work with nurses and upper management to develop a training outline. • Work with individual nurses to train new procedures and order of operations.	Barry, Erin, Andrea, ED nurses Barry, Erin, Andrea	Enforce the new online system and introduce nurses to the practice of charging separately of charting; ingrain the system into new hires.

(*Continued*)

(*Continued*)

Countermeasure	Steps	Responsible Party	Goal
5. Checklist for nursing staff specific to patient disposition to force a self-check at the end of documentation in front of each nurse station.	Discuss with nurses and staff. Develop checklist with high levels of user input. Place checklist at each nurse computer and have nurses begin to self-check documentation.	ED Nurse staff, Andrea ED Nurse Staff, Andrea Andrea	Will cause nurses to error proof their own documentation by checking if they dropped charges after each patient.
6. Weekly audit of charges at restock by Dorian to report on missed charges from the previous day.	Discuss with Dorian the idea of an audit and how often it could be performed.	Dorian, Andrea	Front end progress check to keep nurses in the loop of how they are doing and allow for feedback/continuous system improvement.

Target Condition:

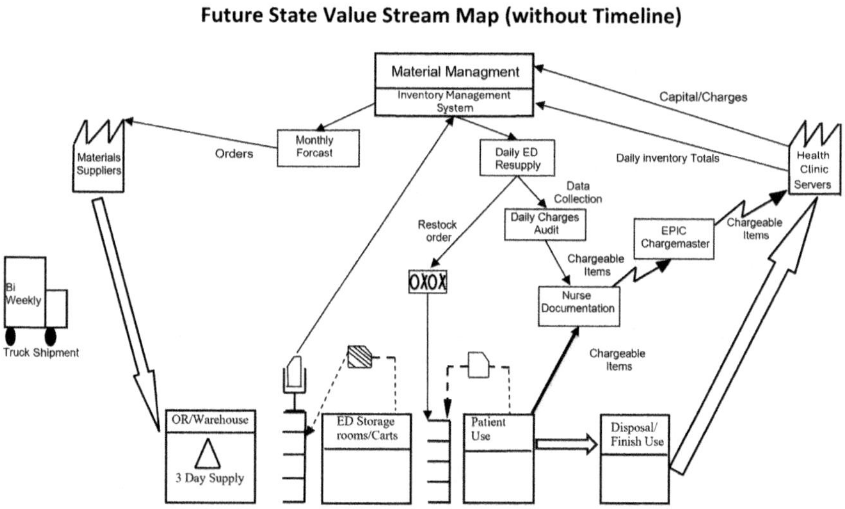

Implementation Plan

Action	Responsible	Deadline	Expected Outcome	Actual Outcome
Standardize Epic format uses by all nurses and discuss with nurses and upper mgmt.	Andrea, Barry, Erin	Friday (week 1)	See Countermeasure #2	Expected outcome met.
Place notice of incoming changes at nurses' station.	Andrea	Friday (week 1)		Notice placed.
Run training for the new charge system and methods of software use.	Andrea, ED Nurses	Monday (week 2)	See Countermeasure #4	Expected outcome met.
Implement a new format in Epic for all nurses.	Andrea, Barry, Erin	Tuesday (week 2)	See Countermeasure #3	Expected outcome met.
Develop and place a checklist at nurses' stations.	Andrea, ED Nurses	Tuesday (week 2)	See Countermeasure #5	Expected outcome met.
Begin use of all online charge system.	Andrea, Barry, Erin, ED Nurses	Tuesday (week 2)		Online charge system in use.
Begin biweekly audit of chargeable materials using inventory management system.	Dorian, Andrea	Thursday (week 2)	See Countermeasure #6	Expected outcome met.

Effects Confirmation

FOR EACH PATIENT:

PLEASE REVIEW IF YOU HAVE COMPLETED THE FOLLOWING TASKS WHEN A PATIENT IS DISCHARGED or THEIR CHART IS CLOSED

- ☐ Vitals Completed
- ☐ Pain assessment completed
- ☐ GCS Completed
- ☐ Lines and Drains Completed (If needed)
- ☐ Follow up/DC completed
- ☐ Charge capture completed
- ☐ Discharge Checklist Completed
- ☐ Disposition Completed
- ☐ Meds and Med Orders Completed (If needed)
- ☐ Discharge instructions given
- ☐ Discharge Patient (In Unit Census)

Training Procedures for Charge Capture Improvement

Overall Goal:
Orient nurses to new online charging procedures and assist them in changing their EPIC screens to a common format to assist with charge reminders. Train staff on the common flow of dispositions and the most-used items.

Most Used Items:

1) *OXI Sensor ADULT*
2) *OXI Sensor NEO/ADULT*
3) *SPLIT CAST PRE-CUT 5 x 30*
4) *ETHILON SUTURE 4-0*
5) *FOLEY CATH TRAY 16 FR KIT*

Actions:

1) The following order in the **discharge section** of the **disposition tab** should be organized in each nurse's EPIC:
   1) Vitals
   2) Pain
   3) GCS
   4) Lines and Drains
   5) Follow Up/DC
   6) Charge Capture
   7) Discharge Checklist
   8) Disposition

   ***Organization beyond #8 is up to the nurse's preference**

2) *Change charge capture format by clicking the wrench, selecting the written description, and changing the number of columns to 3. Accept the changes.*

3) *Go over the laminated list of chargeable items with staff members; highlighted and bolded items are the most consistently used chargeable items.*

4) *Walk through the discharge checklist. Before finishing documentation, all nurses should review key aspects of the discharge section to ensure important documentation is completed and charges were not missed.*

Changes Made During Implementation:

- ❖ Training form was checked over and modified by head nursing staff.
- ❖ Flow of discharge documentation check sheet was modified after gaining staff feedback.
- ❖ To increase reach of training, email was sent out to staff with instructions to cover all bases by head nurses.
- ❖ Former nurse EPIC trainer was able to modify format and names of charges and begin standardizing charge formatting to eliminate unnecessary charges and then copy the format to all other nurses EPIC's.

General Observations:

- ❖ Staff seemed to be accepting and on board with the improvement methods.
- ❖ Staff tended to be more cognizant of charges during the course of the project and observation period.
- ❖ Many nurses had suggestions for general improvements in the ED.

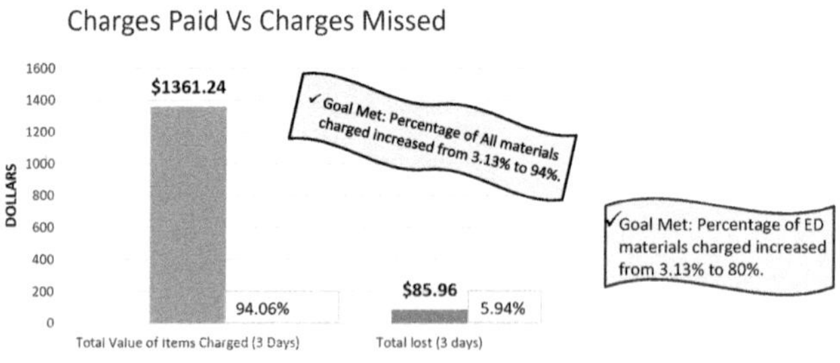

Note: Only Material Charges were included in data collection

Follow-Up Actions

Action	Responsible Party	To Complete Action	Due
Continue weekly audits of charge capture in ED and show numbers to nurses to foster feedback and continuous improvements at least for the next month.	Dorian	Pull inventory data and receive charge capture data, analyze data, and present to nurses and lead staff on a monthly basis.	Ongoing.
Have Fernand make a refined preference list and program in only the most used charges into all other nurses' Epics.	Fernand, Barry, Erin	Create a list of most used items and import into all other nurses' Epics when they are available.	In 1 month.
Recommend hiring more nurse/assistant staffing for ER to allow for delegation of tasks and increase inflow and outflow of patients in the ER as volume increases.	Erin, Barry, Chantal	Contact employee management, interview and hire new staff, and train them in ED procedures.	In 2 months.
Combine charges for similar materials and procedures to create a small set of standard charges in the next Epic update, to allow faster and easier charge capture by staff.	Epic Team, IT, Chantal	Work with Epic accent team to include user (nurse) feedback for ED-specific Epic software to make charge capture easier.	In 3 months.

Case Study 7.4: Capture ER charges more efficiently	Date: XX/XX/20XX
Sponsor: MC, Nursing Department	**Author: Nestor**

Background
• The MC wants an ER Charge Capture System that is:
  • Easy to use without interfering with patient care.
  • Simple to convert to billing.
  • Fully captures ER charges.
• In last 18 months: Average of 4 ER visits per days, and 1 trauma per month.	

Current Condition

ER charge capturing process

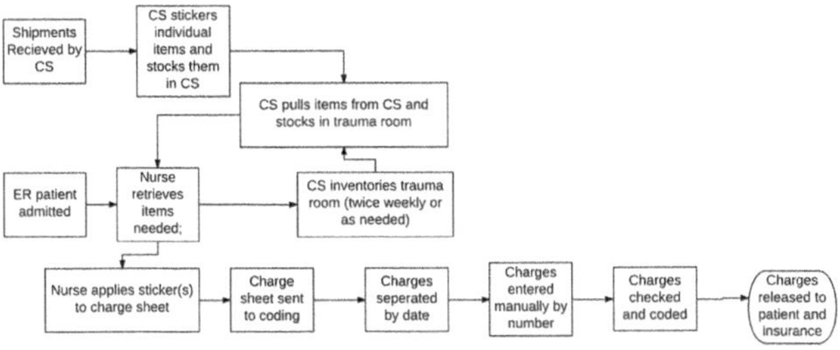

ER charge sheet, Physician page

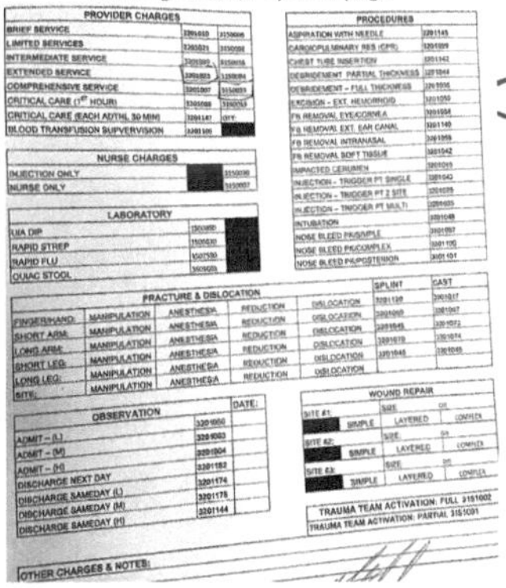

ER charge sheet, Nurse page

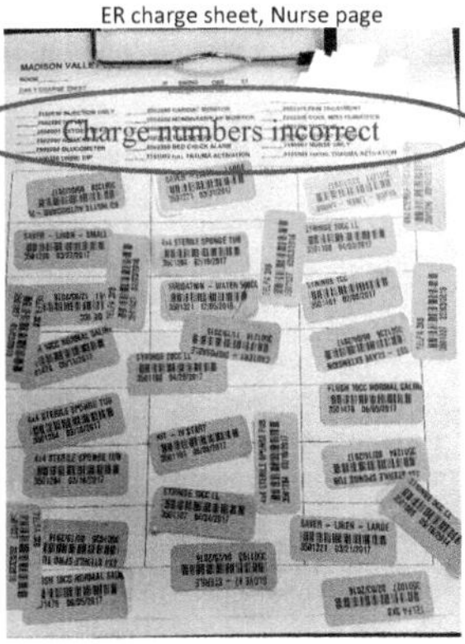

Sticker location during ER case

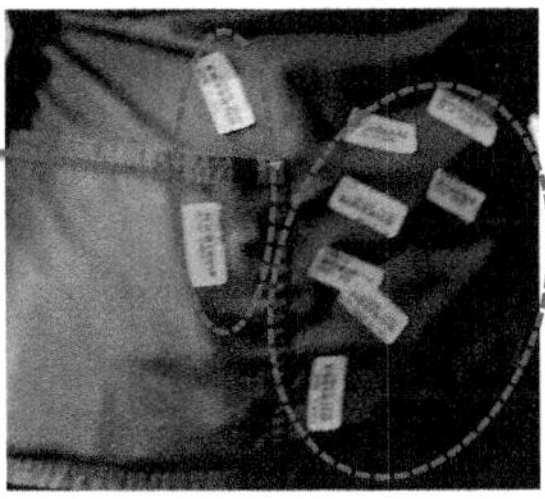

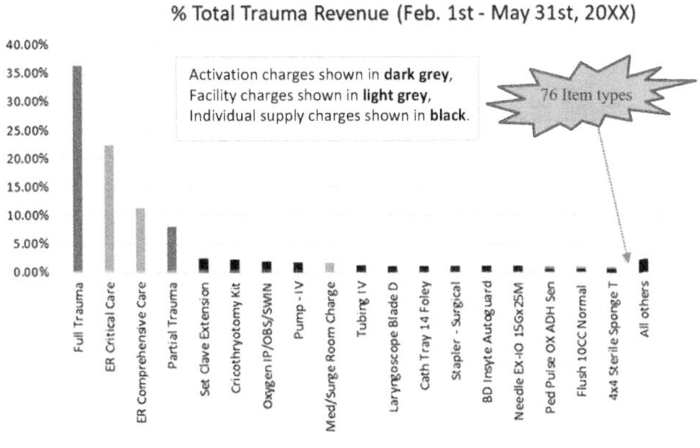

Assumptions:

- *Average profit per patient from supplies = $27.48 (plus or minus $7.75 per patient at 20% error);*
- ***System cost is $3.70 per patient**, so true **average profit per patient from supplies = $23.48***
- *Average profit calculated by subtracting 90% of the billable items account balance from ER Revenues Year to Date (YTD) and dividing by total number of ER and Med Only patients YTD.*

NASA-Task Load Index

- The director of nursing hypothesized that the current charge system was not working because it placed **too much burden on the nurses**, and that one way to capture more charges was by increasing the usability of the charge system for the nurses.
- NASA-Task Load Index was used to capture the **workload** that the current charging system places on the ER nurses.
- Four (4) nurses chose to respond to the assessment as soon as practical after individual ER cases.

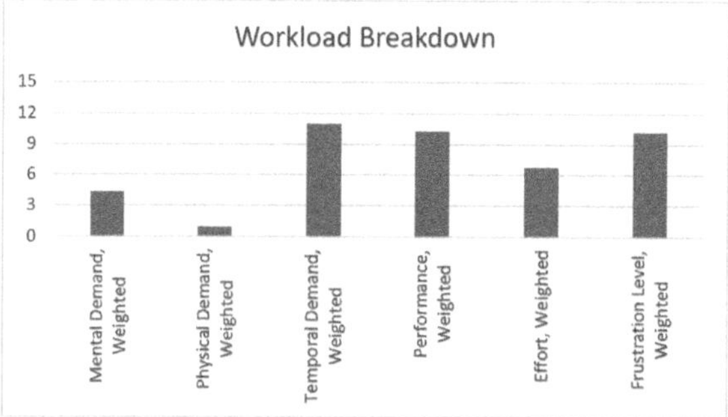

Average Scores for the Six Dimensions of the NASA-TLX

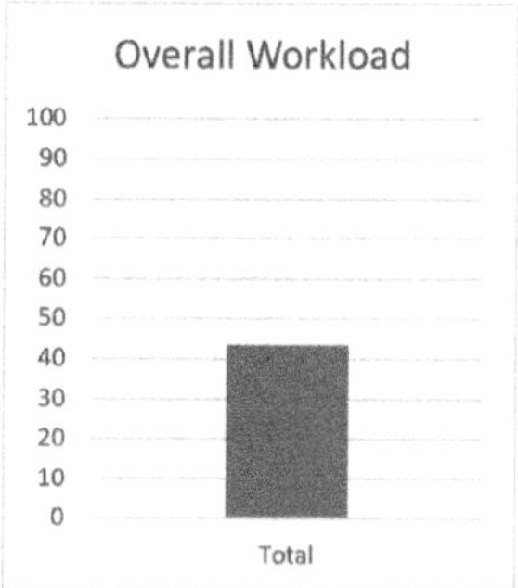

Total NASA-TLX

SMART Goals:	• Decrease System Cost per patient by 30% (from $3.70 down to $2.59 → reduction of $1.11/patient). • Decrease nurses' workload during cases by 15% (Task Load Index decrease from 42 to less than 35.7).

Root Cause Analysis

Cause: Not all charges are being assigned to patients during ER visits

Potential Countermeasures

- Over 100 types of individually stickered items in the ER → • Reduce number of stickered items in the ER.
- Charging is a secondary concern, after patient care → • Relieve nurses from charging duties or integrate better into patient care.
- Coding cannot tell if supply charges have been missed → • Create reconciliation for supplies after ER cases.

Proposed Countermeasures (Course of Actions; COA):
COA1: Replace the sticker system with an itemized charge list; reconcile charge list with nurses' charts at the end of ER cases (no charging during case).
COA2: Develop blanket charges that encase individual line items for coding (one sticker for multiple supply charges); develop a standard reconciliation process for both nurses and coders.
COA3: Remove stickers from "low level" items (suggest <$10, 45 total); increase facility fees (suggest $10 per patient).

COA	Pros	Cons
1.	• Completely eliminates stickers in ER. • Should not increase coding workload. • Eliminates nurse charging during cases. • More opportunities to catch errors.	• Large system change for short term fix. • Initial learning curve may temporarily increase nurses' workload.
2.	• Reduces the number of stickers in the ER. • More opportunities to catch errors.	• Lost stickers will have larger consequences. • Probably not useful after the Electronic Health Record (EHR) system change. • The reconciliation process increases workload for nurses and coders. • Limited opportunity to group items.
3.	• Halves the number of stickers in the ER. • Should decrease nurse workload and coder workload. • Could still be pertinent even after the EHR system change.	• Reimbursement risk associated with facility fees. • Risk associated with error calculating the appropriate increase.

- Moving forward with COA 1 for IV supplies only (adding IV items to the current sheet as line items).

Implementation Plan

Action	Responsible	Deadline	Expected Outcome	Actual Outcome
Revise the charge sheet to include IV items as line items.	Nestor	Monday (week 2)	New charge sheet developed.	Finished, pending nurse feedback for adjustments.
Remove stickers from IV items in the ER.	Nestor	Tuesday (week 2)	No stickers on IV items in ER.	All IV items in ER have had stickers removed by Tuesday.
Replace old charge sheets with the new version.	Nestor	Tuesday (week 2)	All previous charge sheets gathered.	New versions of charge sheet in place in ER and inpatient by Tuesday.
Remove stickers from IV items in Med room.	Nestor	Tuesday (week 2)	No stickers on IV items in Med room.	All IV items in Med room have had stickers removed by Tuesday.
Remove stickers from IV items in Central Supply.	Nestor	Tuesday (week 2)	No stickers on any IV items in Central Supply.	All IV items in Central Supply have had stickers removed by Wednesday.

Effects Confirmation

Revised ER Nurse Charge Sheet

Revised ER Nurse Charge Sheet

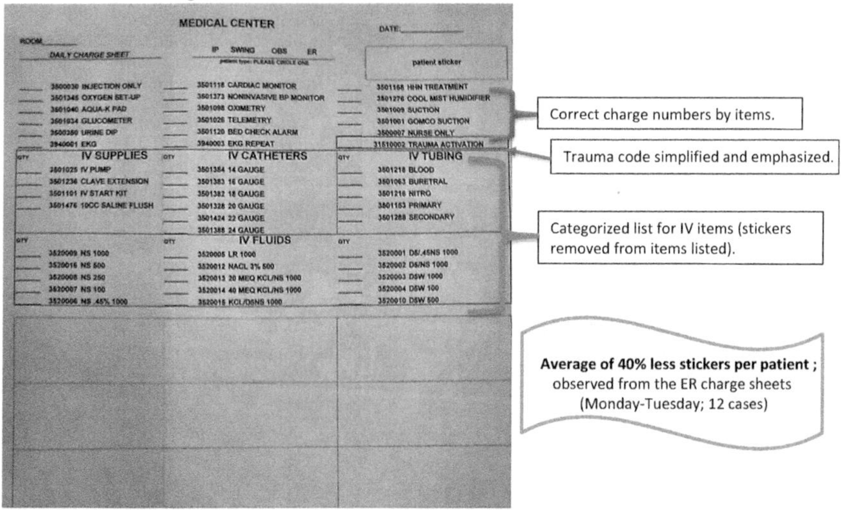

Accomplishments:

- All IV items in ER have had stickers removed.
- New versions of charge sheet in place in ER and inpatient.
- All IV items in Med room have had stickers removed.
- All IV items in Central Supply have had stickers removed.

Administered NASA-TLX to nurses after ER Cases on the New System (data for 6 different nurses):

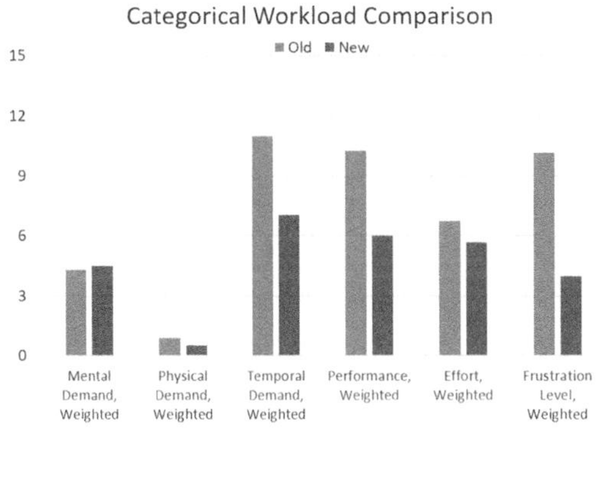

Average Scores for the Six Dimensions of the NASA-TLX Old vs. New System

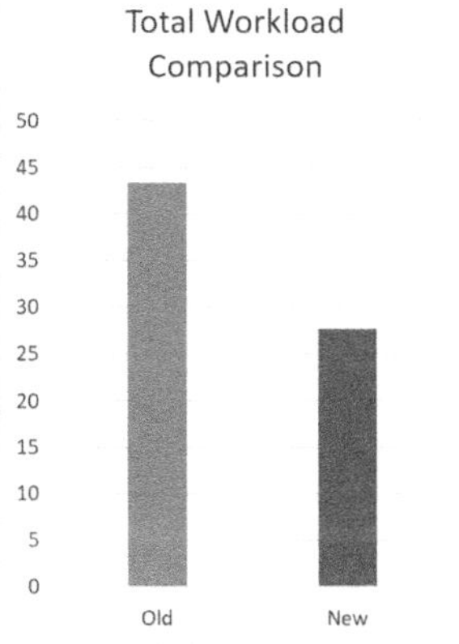

Total NASA-TLX Old vs. New System

Five out of six NASA-TLX dimensions decreased.
Mental demand increased a little due to learning curve of new system.

Total NASA-TLX decreased by 35.7%!

Estimated annual savings for new system *(savings estimated based on breakdown of completed charge sheets)*:

1460 visits/year x $1.23 savings/patient = **$1795.80 annual savings from new system**

- $0.24 from material cost
- $0.43 in purchasing labor
- $0.56 in nursing/coding labor

System Cost per patient decreased by 33.24%!

Year	20XX				20XX+1				20XX+2			
Growth	0%	5%	7.6%	10%	0%	5%	7.6%	10%	0%	5%	7.6%	10%
New System	$1,796	$1,796	$1,796	$1,796	$1,796	$1,886	$1,932	$1,975	$1,796	$1,980	$2,079	$2,173
COA1	$3,358	$3,358	$3,358	$3,358	$3,358	$3,526	$3,613	$3,694	$3,358	$3,702	$3,888	$4,063
COA2	$1,621	$1,621	$1,621	$1,621	$1,621	$1,702	$1,744	$1,783	$1,621	$1,787	$1,876	$1,961
COA3	$2,701	$2,701	$2,701	$2,701	$2,701	$2,836	$2,906	$2,971	$2,701	$2,978	$3,127	$3,268

Year	20XX+3				20XX+4			
Growth	0%	5%	7.6%	10%	0%	5%	7.6%	10%
New System	$1,796	$2,079	$2,237	$2,390	$1,796	$2,183	$2,407	$2,629
COA1	$3,358	$3,887	$4,183	$4,469	$3,358	$4,082	$4,501	$4,916
COA2	$1,621	$1,876	$2,019	$2,157	$1,621	$1,970	$2,172	$2,373
COA3	$2,701	$3,127	$3,365	$3,595	$2,701	$3,283	$3,621	$3,955

∗Table represents a rough forecast of the cost savings of each proposed course of action, given different increases in ER visits. Labor savings for COA1 and 2 are mostly speculative. Table does not account for risk of unrealized or uncaptured charge, which varies between countermeasures.

Follow-Up Actions

Action	Responsible Party	To Complete Action	Due
Track ER supply usage.	Olga	Keep log of supplies ordered for ER with the provided spreadsheet.	In 3 months.
Post reminders for nurses.	Olga	Choose location and post reminder signs.	In 3 months.
Evaluate capture effectiveness of new charge sheets.	Olga	Calculate profit per patient using July and August costs, revenues, and number of patients.	In 3 months.
Plan for materials management and charge capture process with a new EMR system.	Olga	Determine how materials will be tracked from order until delivery to the patient (paper, scanners, etc.) and take appropriate steps to prepare for new EMR.	In 3 months.

Case Study 7.5: Increase Accuracy and Efficiency in the Charge Capture Process for ER	Date: XX/XX/20XX
Sponsor: ED	**Author: Pablo**

Background

- The current patient charting process is cumbersome and repetitive, leading nurses and doctors to mistakenly omit important information or only include it in part of the patient chart.
- This can cause the coders to miss important information regarding what went on in the case and what can and should be charged to the patient or their insurance.
- There is currently no method to verify that all of the supplies used and procedures or assessments/diagnoses for a particular case were accounted for in the patient chart or billing statement.

Current Condition

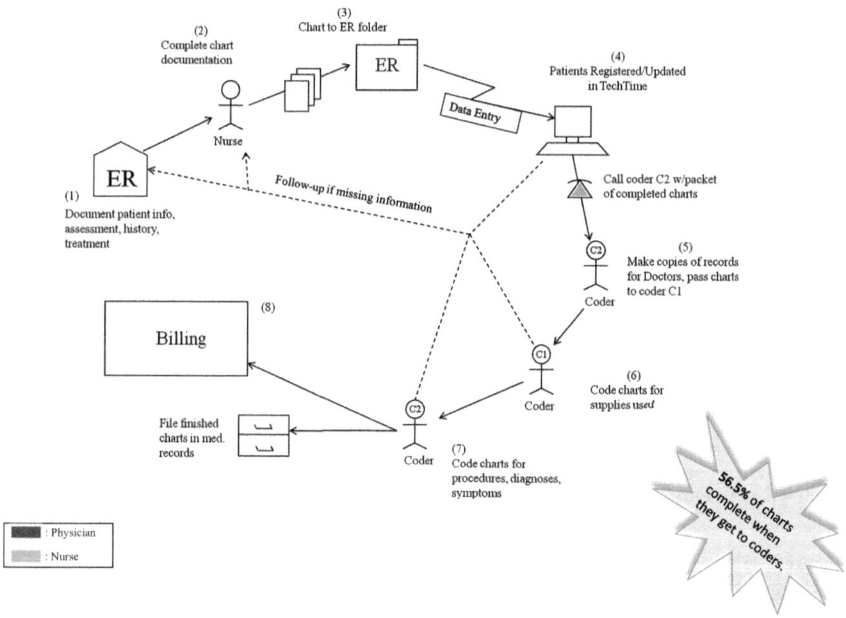

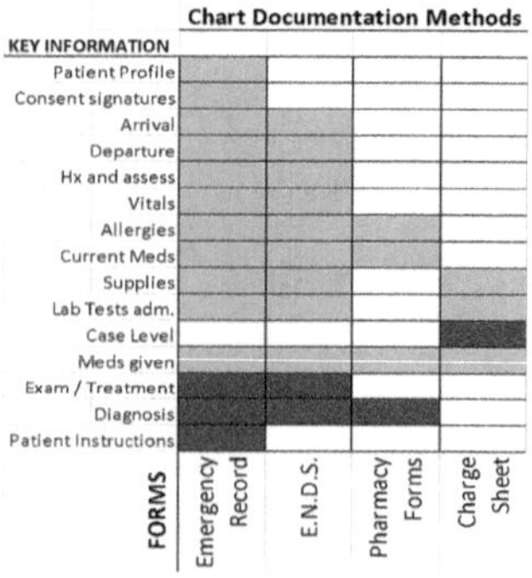

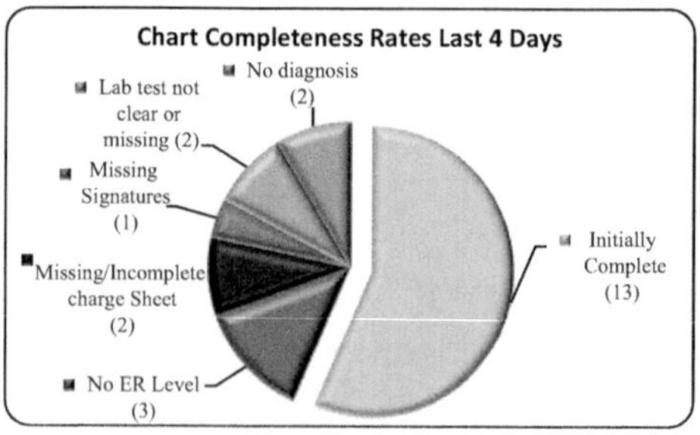

SMART Goal:	• **Increase the amount of charts completed upon arrival in coding to at least 75%.**

Root Cause Analysis

Reasons for delays or inaccurate/inadequate information in the charting/charge capture process:

- **Patient personal profile information not complete.**
  - **WHY?** Nurse mistakenly omitted info or signature.
    - **WHY?** Chart packet too cumbersome to finish for quick or intensive cases.
- **Chart could not be coded for supply charges.**
  - **WHY?** Missing charges.
    - **WHY?** On charge sheet, but not on record.
      - **WHY?** Charts require repetitive documentation.
    - **WHY?** On record, but not on the charge sheet.
      - **WHY?** Supply may not have charge inventory stickers.
    - **WHY?** Specific type or amount not provided.
      - **WHY?** The nurse is unaware this is required.
        - **WHY?** Not specific on the charge sheet.
  - **WHY?** Missing doctor signature.
  - **WHY?** Missing patient signature.
    - **WHY?** The nurse forgot to get it.
  - **WHY?** Missing case level.
    - **WHY?** The doctor did not look at the charge sheet.
      - **WHY?** The nurse is responsible for charges, and the case level is recorded on the charge sheet.
- **Chart could not be coded for diagnosis.**
  - **WHY?** Information missing.
    - **WHY?** Not enough room for the nurse/doctor to write everything.
    - **WHY?** Inconsistent on both sheets.
  - **WHY?** The procedure used is not supported by symptoms/diagnosis.
    - **WHY?** The doctor may have been trying to eliminate diagnoses with lab tests.
    - **WHY?** The nurse or doctor may have omitted symptoms or signs.
      - **WHY?** Not enough room to fit everything.
      - **WHY?** May have been rushed or busy with other duties.

Proposed Countermeasures

1. Reformat the Emergency Record sheet to be used only for patient personal information/consent signatures and the doctor's assessment and treatment of the patient as well as diagnosis.
   a. The case level will also be recorded in this area to be convenient for the doctor.
2. Eliminate the charge sheet by including a place to put inventory stickers on the revised Emergency Record sheet so that nurses can more easily record charges as the supplies are used.
3. Implement inventory stickers that are more easily removed and implement for all supplies.

Target Condition:

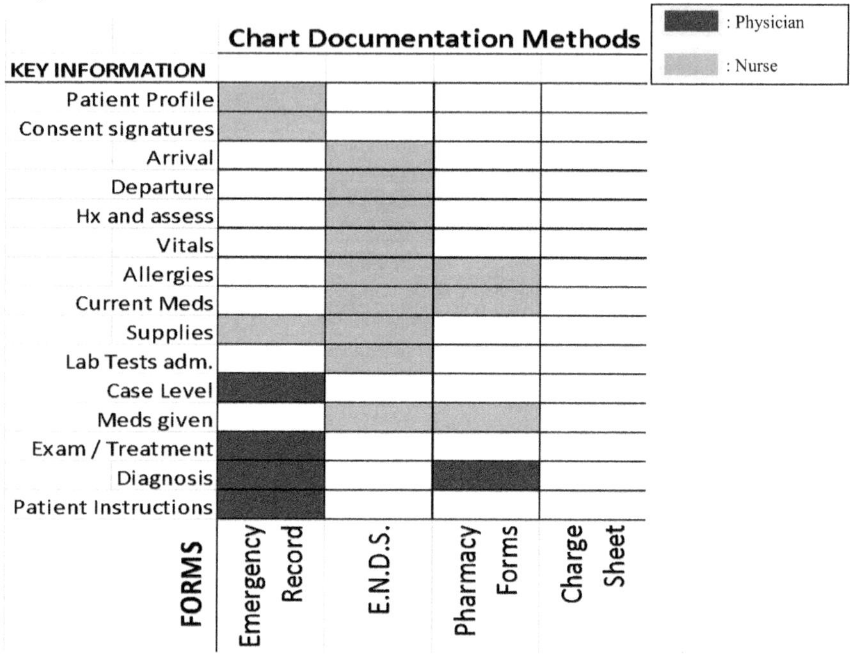

Implementation Plan

Action	Responsible	Deadline	Expected Outcome	Actual Outcome
Propose revised chart to ER nurses and coders.	Sebastien	Monday (week 2)	All nurses and coders aware of the pilot plan.	Expected outcome met.
Propose revised chart to physicians.	Sebastien	Monday (week 2)	All doctors aware of the pilot plan.	Expected outcome met.
Obtain feedback from staff and physicians and make improvements to the pilot chart accordingly.	Sebastien	Tuesday–Wednesday (week 2)	Make a new chart optimal for all parties involved.	Expected outcome met.
Answer any staff or physician questions regarding changes.	Sebastien	Tuesday–Wednesday (week 2)	Questions and concerns resolved.	Expected outcome met.
Print out trial chart copies and replace old ER sheets.	Sebastien	Tuesday–Wednesday (week 2)	Allow staff to use the proposed new chart.	Expected outcome met.
Begin pilot study with a trial chart.	All staff and physicians	Tuesday–Wednesday (week 2)	The pilot study begins.	Expected outcome met.

(*Continued*)

(Continued)

Action	Responsible	Deadline	Expected Outcome	Actual Outcome
Data collection on the number of charts that can be coded without delay during pilot study.	Sebastien	Wednesday–Friday (week 2)	Data collected to test the significance of improvement.	Expected outcome met.
Present final report and follow-up actions.	Sebastien	Friday (week 2)	The pilot study is either continued or not.	The pilot study continued.

Effects Confirmation

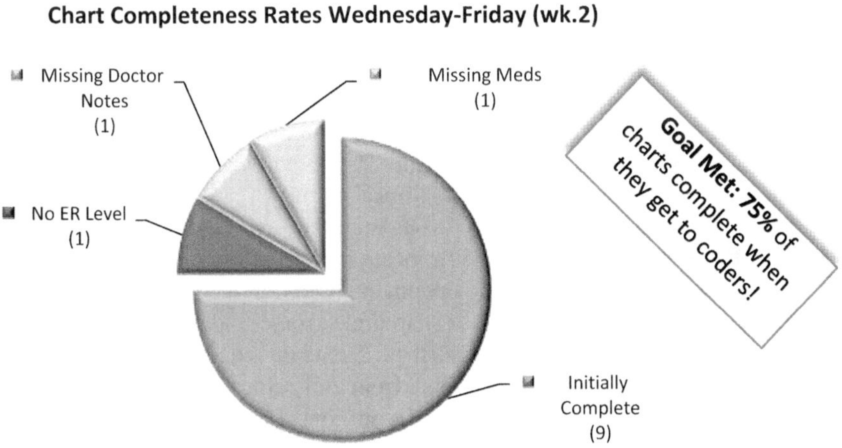

Follow-Up Actions

Action	Responsible Party	To Complete Action	Due
Continue documenting instances when charts are missing important information for the coders.	All staff and physicians	Follow action	Ongoing.
Implement the new stickers that can be peeled off and stuck onto the revised ER sheet for all supplies.	All staff and physicians	Follow action	In 1 month.
Use visual cues to remind nurses and physicians to use the sticker system once it is implemented.	All staff and physicians	Follow action	In 1 month.
Use root cause analysis for these instances and brainstorm/test ideas to eliminate the sources of inefficiency or inaccuracy.	All staff and physicians	Follow action	In 3 months.

8 Case Studies Related to Workflow in the Business Office, Human Resources, Lab, and Dietary Services

In this chapter, RIE case examples focus on improving administrative efficiency and operational flow across the Business Office, Human Resources, Laboratory Services, and Ancillary Services. The RIE briefs demonstrate how the A3 thinking process and the PDSA framework can simplify processes, enhance communication, and eliminate waste in both clinical and non-clinical settings.

Case 8.1 addressed the challenge of outdated and excessive personnel policies in Human Resources. Staff struggled to locate policies, taking over 2 minutes per search and often opening the wrong document. Through a 5S process, redundant and obsolete policies were removed, a shared electronic folder was created, and a standardized review schedule was established. The team reduced the number of personnel policies by 44% and cut search time from 2.11 minutes to just 19.9 seconds.

Case 8.2 focused on patient privacy and workflow efficiency in a lab constrained by space and layout. By removing one of two drawing chairs, reorganizing the room for single-patient privacy, and introducing a shared electronic filing system, the team improved patient privacy from 43% to 100% and decreased staff walking distance by 38%. The result was greater efficiency, better ergonomics, and higher patient satisfaction. A key takeaway was that sometimes less is more; having one phlebotomist with a balanced workload proved more efficient than two working in parallel.

Case 8.3 examined front-desk workflow and patient privacy in the Patient Accounts Office. The space allowed conversations to be overheard and patients to bypass check-in. The office was redesigned with new desk arrangements, partitions, and a relocated waiting area. White noise machines, fans, and plants were added to reduce sound transmission. Following these changes, no patients bypassed the front desks, and staff rated improvements in privacy and workflow between 4.4 and 4.7 out of 5.

Case 8.4 streamlined the clinic's billing process, which previously took an average of 6.5 days to complete. The team standardized charge ticket sorting, established regular retrieval schedules, and improved communication between the Health Information Management and billing staff. As a result, processing time decreased, documentation improved, and charge capture accuracy increased.

DOI: 10.1201/9781003670698-8

Case 8.5 reduced unnecessary patient waiting between clinic visits and follow-up lab or radiology services. The team introduced location-coded stickers to improve communication among registration, nursing, and ancillary staff, enabling lab and radiology personnel to identify patient locations immediately. This eliminated unnecessary waiting and clarified responsibilities among departments.

Across these five RIEs, a common theme emerges: healthcare efficiency extends beyond clinical procedures. Administrative processes, communication systems, and physical layouts all play crucial roles in patient experience and staff productivity. By standardizing workflows, enhancing visibility, and promoting accountability, these teams demonstrated that RIEs can yield meaningful results in every corner of a healthcare organization.

Case Study 8.1: Personnel Policy Reduction and Standardization	Date: XX/XX/20XX
Sponsor: Human Resources	**Author: Bonnie**

Background

- Los Medicos Center's Human Resources Department is responsible for informing approximately 110 employees of all personnel-related policies that may affect their day-to-day work lives.
- These personnel policies have not been revised on an up-to-date system in over 20 years, resulting in an excessive amount of personnel policies.

Current Condition

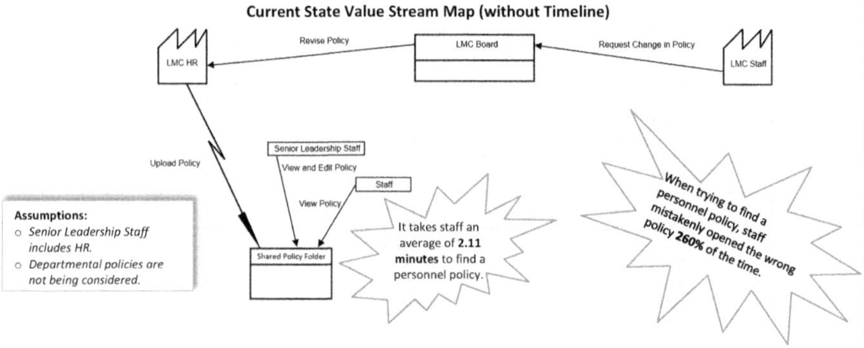

Major Problems:

- There is an excessive amount of time spent on trying to find personnel policies.
- The policies are not revisited or revised on a scheduled basis, so the policies lack proper signatures and updates.
- The policies are redundant and lack clear naming.
- 328 pages of personnel policies are in a binder that is not up to date.
- Attachments of policies are not filed electronically.
- Within the personnel policy electronic folder, there are many policies that pertain to departments and not personnel.
- Many policies contain procedures on how to carry out the policy.

SMART Goals:	
	• **Reduce the amount of personnel policies by 10%, from 161 policies to 145 or less.** • **Reduce the time it takes for employees to find policies to under 1 minute.**

Root Cause Analysis

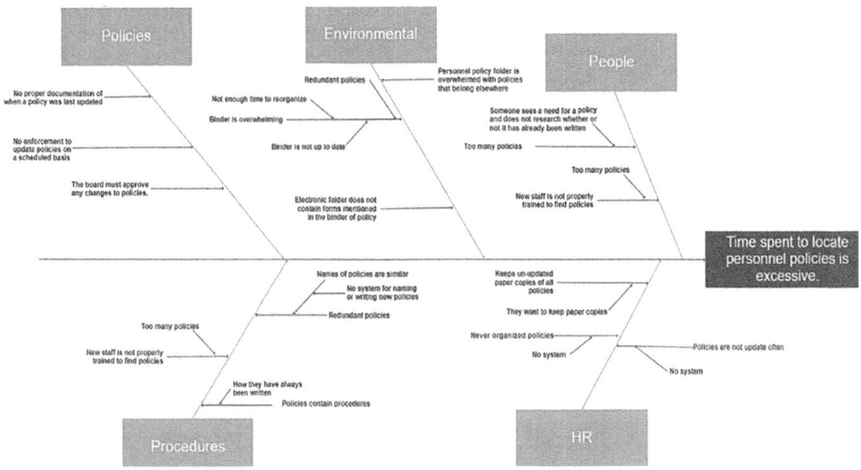

5 Whys:

Problem Statement 1: Takes too long to find personnel policies.

↳ **Why?** The electronic folder of policies is unorganized.

↳ **Why?** There is an excess amount of policies.

↳ **Why?** Old policies are not archived.

↳ **Why?** No system to hold anyone accountable for organization and updates of the policies.

Problem Statement 2: Takes too long to find personnel policies.

↳ **Why?** The electronic folder of policies is unorganized.

↳ **Why?** There is an excess amount of policies.

↳ **Why?** Redundant policies are not combined.

↳ **Why?** No system to hold anyone accountable for organization and updates of the policies.

Problem Statement 3: Takes too long to find personnel policies.

↳ **Why?** Policies are hard to identify/find in the electronic folder.

↳ **Why?** The names of the policies are similar.

↳ **Why?** Different policies are redundant and have similar meanings.

↳ **Why?** Old policies were never combined or archived.

↳ **Why?** No system to hold anyone accountable for organization and updates of the policies.

Problem Statement 4: Too many personnel policies.

↳ **Why?** When creating new policies, no one checked to see if they already had a policy for that topic.

↳ **Why?** No proper documentation of current policies.

↳ **Why?** The policies are not revised/revisited on a scheduled basis.

↳ **Why?** No system to hold anyone accountable for organization and updates of the policies.

Proposed Countermeasures

Counter Measures	Needed Actions	Responsible	Expected Outcome
1. 5S personnel policies.	• Sort all personnel policies into categories. • Move any policies that do not fit into a category to another department or archive them. • Create a new shared folder of personnel policies. • Combine any redundant policies into one.	Bonnie and Alex Bonnie and Alex Bonnie and Colin Bonnie and Alex	• Personnel policies will be easy to find. • A reduction in personnel policies. • Policies will be up to date. • Policies will be where they need to.
2. Add a monthly calendar reminder for revisits of personnel policies.	• Determine best times to review policies. • Determine what categories should be reviewed when, so that every policy is reviewed annually. • Determine who is responsible for reviewing policies annually. • Add a monthly reminder to the responsible party's calendar.	Bonnie and Alex Bonnie and Alex Bonnie and Alex Bonnie and Colin	• Every policy will be reviewed annually. • Policies will remain up to date. • They will become more familiar with the policies, so they are less likely to create redundant ones. • Eliminates need to create policies randomly.
3. Policy on policies flowchart.	• Create a flowchart on how to make a policy.	Bonnie	• Allows people to see whether or not they actually need to create a policy. • Staff won't request new policies all the time.
4. Policy template.	• Create a policy template.	Bonnie	• Policies will be in the same format. • Policies will be up to date.
5. System for documenting approved and revisited policies.	• Create a sheet for signing in the binder.	Bonnie	• Keep track of who edits a policy. • Hold people accountable.

Target Condition:

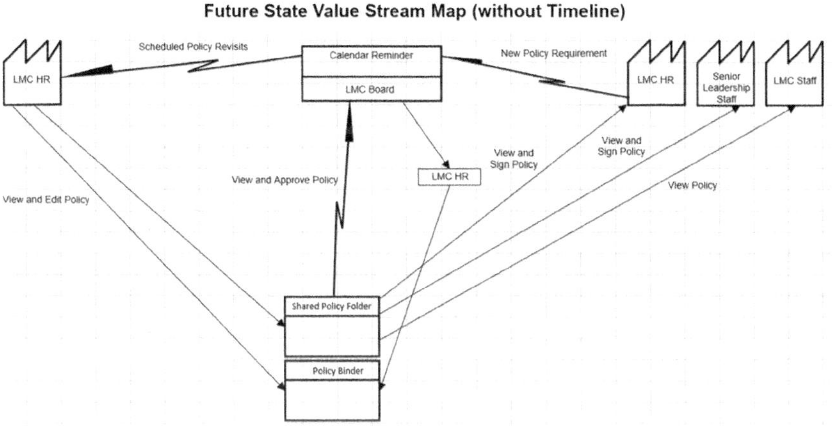

Implementation Plan

Action	Responsible	Deadline	Expected Outcome	Actual Outcome
Organize and categorize all current personnel policies.	Bonnie and Alex	Tuesday (week 2)	See Countermeasure #1	Expected outcomes met.
Organize binder with new policies and eliminate old policies.	Bonnie and Alex	Tuesday–Wednesday (week 2)	See Countermeasure #1	Expected outcomes met.
Inform all staff of any changes made.	Bonnie, CEO, Human Resources, and Senior Leadership Staff	Wednesday–Thursday (week 2)	See Countermeasures #3–5	Expected outcomes met.
Inform Alex of the new calendar reminder.	Bonnie	Wednesday (week 2)	See Countermeasure #2	Expected outcomes met.
Tell Senior Leadership Staff about monthly meetings on policies.	Bonnie	Thursday (week 2)	See Countermeasures #3–5	Expected outcomes met.

Effects Confirmation

5S Methodology

- **Sort** (Separate and remove clutter and items unneeded in the workspace).
  - Removed old, unneeded, and redundant policies.
- **Set in Order** or Straighten/Stabilize (organize what is left to minimize movement and make things clear).
  - Organized policies into categories.
- **Shine**/Sweep/Systematic Cleaning (keep the workplace tidy and organized).
  - Created a policy creation flowchart.
  - Reduced personnel policies by 44%.

- **Standardize** (work practices should be consistent and standardized).
  - Step by step system procedures are in the recommendations document.
- **Sustain** (maintain and review standards).
  - Created a calendar reminder to ensure that the policies are being revisited bimonthly.

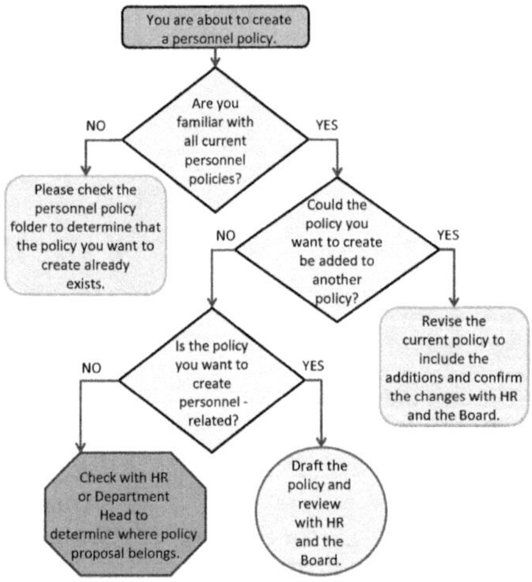

Please fill out this table annually.
*Sign and date "category" rows if and only if the entire category has been reviewed.

Policy	Revised/Revisited By	Date
Category: Benefits		
Employee Assistance Program		
Extended Illness Bank		
Health Insurance		
Holidays		
PTO		
Recruitment Bonus		
Retirement 401K		
Shared PTO		
Sign-on bonus for critical positions		
Category: Breaks and Meals		
Break Periods		

Los Medicos Center Human Resources Personnel Policy:	Policy Name
	Approved by:
Date Revisited/Revised:	Created by:

APPLIES TO:
<u>Who</u> does the policy apply to.

PURPOSE/OVERVIEW:
One or two sentences on **<u>what</u>** the policy covers.

POLICY:
An in-depth explanation of **<u>why</u>** the policy exists and **<u>what</u>** it is.

PROCEDURE:
Explanation of **<u>how</u>** to carry out the policy.

*SHOULD ONLY BE USED ON AN AS-NEEDED BASIS.

Goals Met:
- ✓ Reduced the number of policies by 44%.
- ✓ Reduced search time from 2.11 minutes (126.6 seconds) to 19.9 seconds.

Follow-Up Actions

Action	Responsible Party	To Complete Action	Due
Have staff follow the recommendation document.	Alex, Danielle, Laura, and Earl	Follow action.	Ongoing.
Have Alex, Danielle, Laura, and Earl inform all staff of any changes made.	Alex, Danielle, Laura, and Earl	Follow action.	In 1 week.

Case Study 8.2: Phlebotomy Process Optimization	Date: XX/XX/20XX
Sponsor: Lab Services	**Authors: Ian and Lisa**

Background

- Stuvwxyz Health Center is a critical access hospital that serves the surrounding community and a large subset of the rural population in the neighboring state.
- The hospital has experienced an increased number of patients, and they are experiencing a staff shortage in the lab.
- The current layout of the lab only allows two patients to have blood drawn, with both being able to see and hear each other during the process.
- Further, there is a large glass window that gives other patients a view into the lab.
- Also, to access essential lab and administrative equipment, the doors within the lab space are left open.

Current Condition

Current State Value Stream Map (without Timeline)

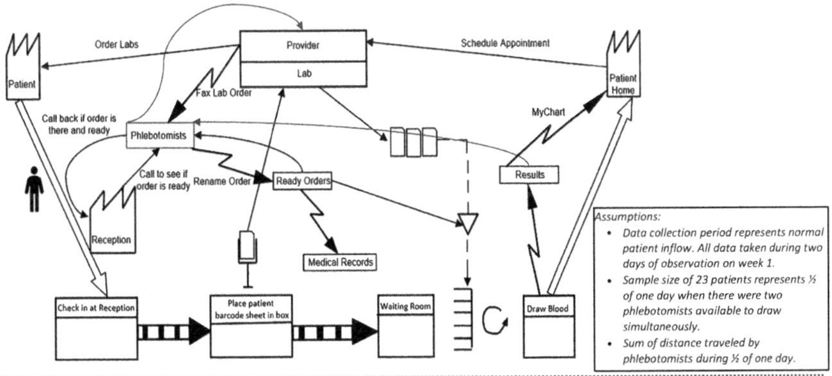

Spaghetti Diagram of Phlebotomist Movement

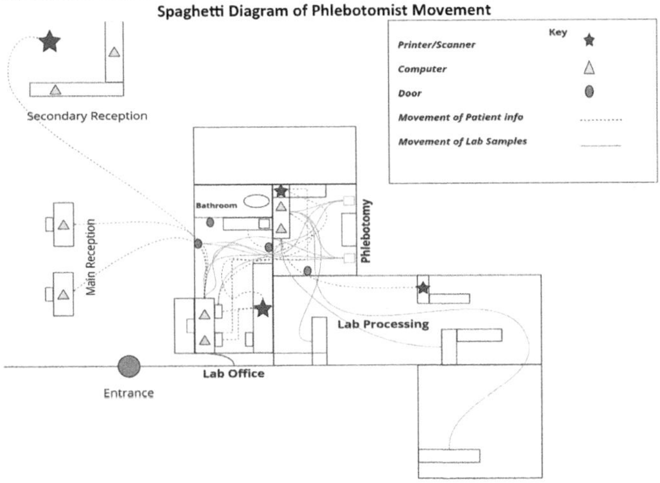

Current Condition

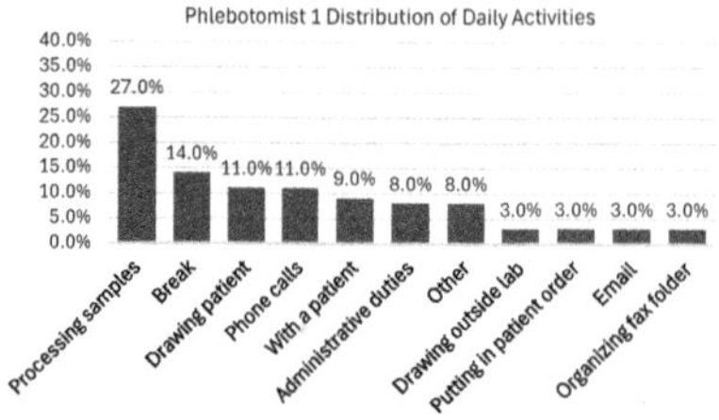

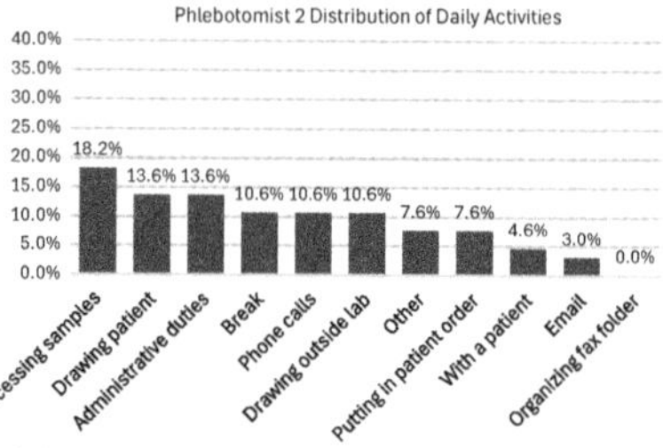

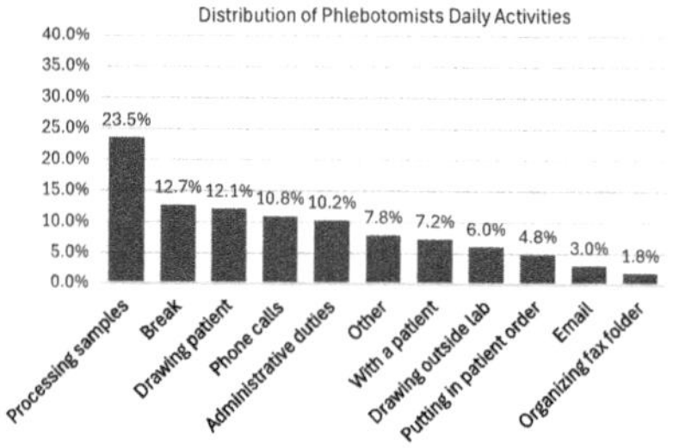

Key Statistics:

- **Of patients who have their blood drawn, only 43% of them have it done in private.**
- **Only 12.05% of a phlebotomist's daily duties involve drawing blood in the lab.**
- **Phlebotomists spend approximately 7.83% of their time doing extraneous activities.**
- **Phlebotomists walk an average of 1,370 ft/hour per shift meaning each of them walks about 2 miles/day.**

Major Problems:

- The level of privacy during blood draws is inadequate for both visual and auditory patient privacy.
- The current organization of the lab space and phlebotomist duties result in a large amount of excess time spent walking when performing essential lab duties.

SMART Goals:	• **Increase the percentage of patients drawn in private from 43% to greater than 90%.** • **Decrease phlebotomists walking by 20%.**

Root Cause Analysis

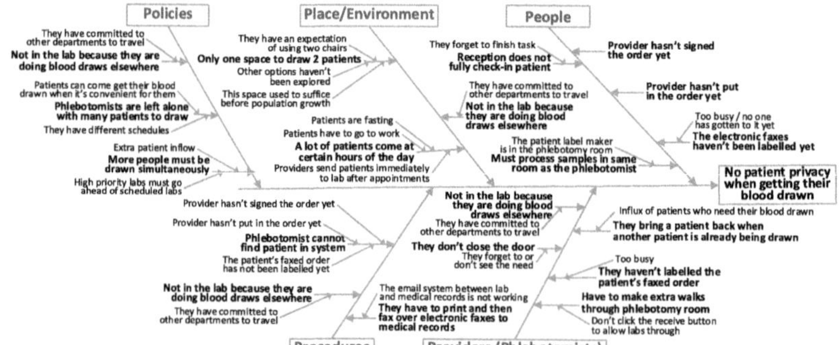

5 Whys:

Problem Statement 1: Patients do not have their blood drawn privately.

↳ **Why?** Patients can hear and/or see another patient getting their blood drawn.
↳ **Why?** There is no separation of patients.
↳ **Why?** There are two chairs in the drawing room.
↳ **Why?** Having two chairs in the same space is seen as being the most efficient use of time.
↳ **Why?** They could see two patients at the same time.

Problem Statement 2: Patients do not have their blood drawn privately.

↳ **Why?** Phlebotomists must move in and out of the drawing room to perform essential duties.
↳ **Why?** Phlebotomists constantly process samples and deliver them to other lab staff.
↳ **Why?** Phlebotomists are overwhelmed with incoming lab orders.
↳ **Why?** Phlebotomists are not efficient enough to keep up.
↳ **Why?** Phlebotomy organization and operating procedures are disorganized or lacking.

Problem Statement 3: Electronic faxed orders take too much time to process.

↳ **Why?** Phlebotomists must print and then fax these orders to send to medical records.
↳ **Why?** Currently, there is no way to share faxes electronically.
↳ **Why?** Their current email system is not working, and they do not have a shared folder for the files.
↳ **Why?** Lack of time to explore alternative options.

Proposed Countermeasures

Counter Measures	Needed Actions	Responsible	Expected Outcome
1. Filing send out labs electronically.	• Inform lab staff of ways to file on their computers.	Ian	• In 2 months, most of their filing cabinet will be empty. • Printer ink and paper not wasted. • Time, energy, and office space not wasted on filing.
2. Creating a shared drive for lab, medical records, and admissions to access patient order forms.	• Have IT create a shared folder with all parties involved. • Inform all departments involved about the change. • Train admissions on how to rename faxes.	IT Lisa and Ian Lisa, Ian, IT	• Reduce the amount of phone calls into lab. • Create an uninterrupted workflow for all departments. • Reduce the amount of work on lab staff.
3. Remove one drawing chair.	• Remove and store one chair.	Lisa, Ian, Julia	• Creates space for the fainting chair. • Enforces lab staff to only draw one patient at a time. • Private drawing for every patient.

(*Continued*)

(*Continued*)

Counter Measures	Needed Actions	Responsible	Expected Outcome
4. Add the fainting chair.	• Find and place the fainting chair in the drawing room.	Lisa, Ian, Julia	• Provides comfort, safety, and accessibility to patients.
5. Create a series workflow.	• Create a flowchart of duties.	Lisa and Ian	• Will allow for all patients to be drawn in private. • Evenly distribute phlebotomist duties.
	• Rearrange workspace for efficient workflow.	Lisa and Ian	• Allow for a different work schedule. • Eliminates need for parallel workflow.

Target Condition:

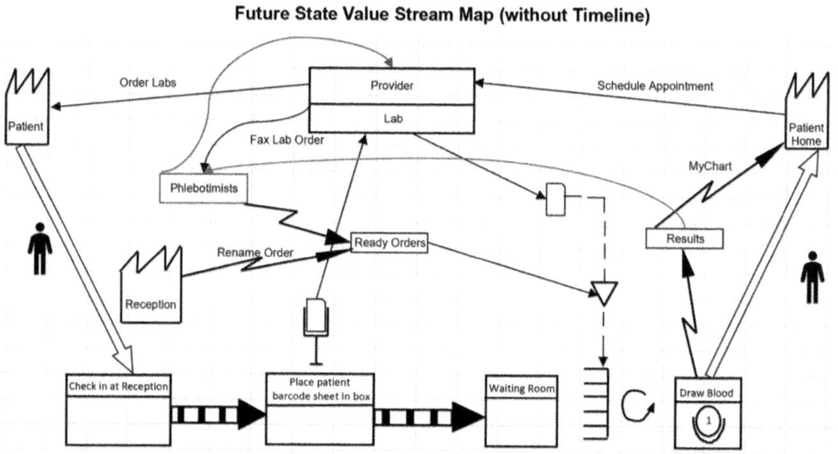

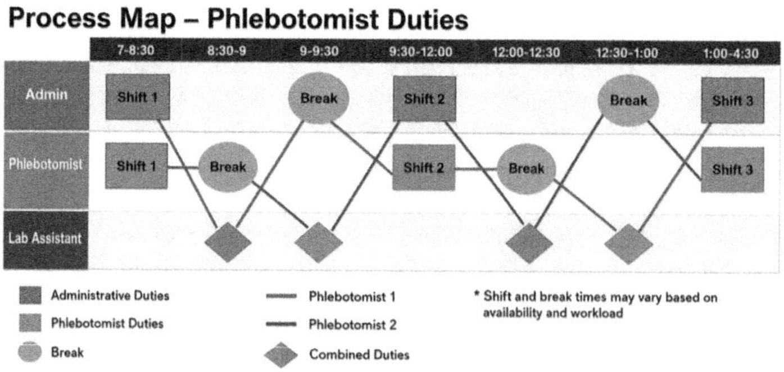

Implementation Plan

Action	Responsible	Deadline	Expected Outcome	Actual Outcome
Have IT create a shared drive with the lab, Medical Records, and Admissions for faxes and train staff on how to access it.	Lisa, Ian, IT	Monday (week 2)	See Countermeasures #1 and 2.	Expected outcomes met.
Rearrange the drawing room with one chair and move drawing materials.	Lisa, Ian, Karl, Julia	Monday (week 2)	See Countermeasure #3.	Expected outcomes met.
Create a one chair system in the lab with split alternating task flows for each phlebotomist. Provide flow charts for each phlebotomist with given tasks for daily rotations. Also include alternate tasks should staffing change.	Lisa, Ian, Karl, Julia	Monday (week 2)	See Countermeasures #3, 4, and 5.	Expected outcomes met.
Brief all lab staff on the new system and guide them on a walkthrough of the system. Assist in any difficulties adapting to the changes and shadowing the first day of work.	Lisa, Ian, Karl, Julia	Tuesday (week 2)	See Countermeasure #5.	Expected outcomes met.
Observe changes and check-in with staff on progress of the new system. Discuss any benefits or negative costs of the system.	Lisa and Ian	Wednesday (week 2)	Adjust the new system.	Expected outcomes met.
Retake all data taken during the planning phase and record differences between new and old data to measure the impact of changes made.	Lisa and Ian	Thursday (week 2)	Record differences between new and old data to measure the impact of changes made.	Expected outcomes met.

Effects Confirmation

Changes Made During Implementation:

- Removed the spare chair from the drawing room. The room now holds one patient at a time.
- Removed fainting chair, to support phlebotomist ergonomic comfort.
- Reinforced the use of an entering patient as a signal to register the next.
- Admin should assist in processing samples when not busy with admin work.

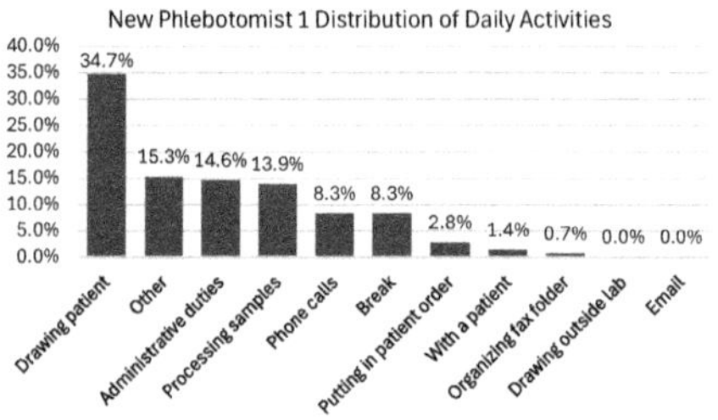

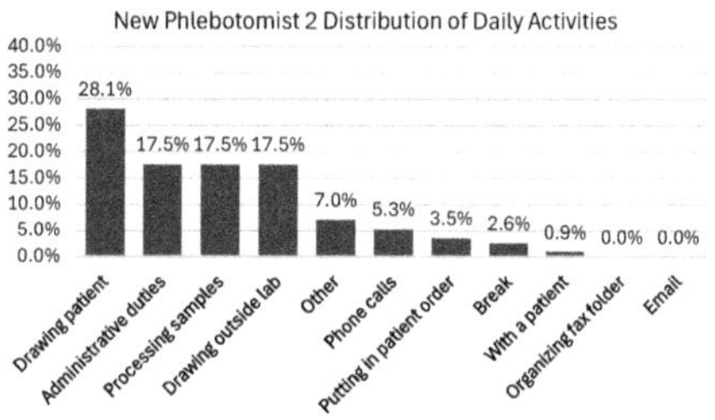

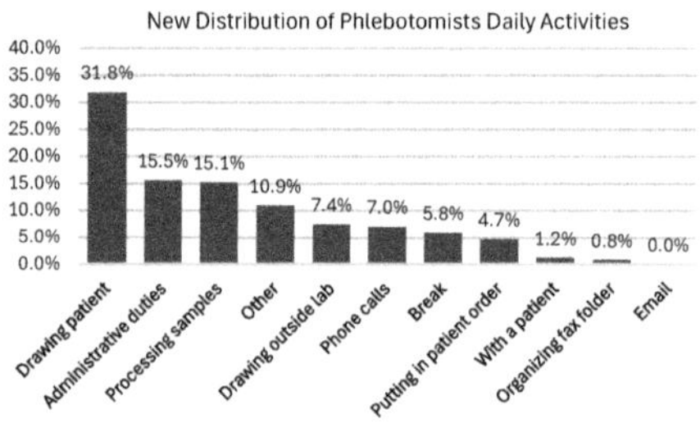

Note: Data collected over 1.5 days,with the assumption that this was a normal workload.

General Observations:

- Staff found that performing draws continuously was taxing; more rotations may prevent excess work fatigue.
- Staff changed procedures to suit their needs with changing workload while still working in series.
- **Overall they felt the workload was less than the old system.**
- Staff were surprised they were able to do the same number of patients with less effort and walking while removing one chair from the drawing room. One of them mentioned: "indeed, sometimes less is more."
  - This is an example that a standardized workflow where tasks are equally divided and assigned can introduce efficiencies in a system.
- **Increased drawing patients from 12.05% with a two-phlebotomist system to 31.78% for a one-phlebotomist system.**

Goals Met:

- ✓ **Increased patient privacy from 43% to 100%.**
- ✓ **Decreased distance traveled per phlebotomist by 38%.**

Follow-Up Actions

Action	Responsible Party	To Complete Action	Due
Order a new adjustable phlebotomy chair for the lab. Order more chairs if needed for staffing requirements.	Matthew, materials, and lab assistants	Order new chair(s) and place them in the lab. Brief phlebotomists on how to use the new chair.	In 2 weeks.
Train any new phlebotomists on the series workflow system.	Karl and Julia	Reference swimlane flowchart and the split workflow description sheet for training.	In 2 weeks.
Once new staff are fully trained, current phlebotomists can rotate schedules to perform lab assistant duties.	Karl and Julia	Always have two phlebotomists to perform blood draws and administrative work. If there are at least two phlebotomists, then one lab assistant can go back to the lab to do their lab work.	In 2 weeks.
Have the manager keep all documents for referencing and updating in the future to ensure patient privacy.	Nicole, Otto, Fiona, and/or new manager	Give Nicole access to all documents created.	In 2 weeks.

Case Study 8.3: Improving Patient Flow and Privacy in Patient Accounts Office	Date: XX/XX/20XX
Sponsor: Business Office	Author: Ana

Background

- Patient Accounts office is in repurposed operating room suite.
  - Patients enter main office – repurposed prep room.
  - Insurance issues handled in two back rooms – entrances from main office back right and left.
- Staff are concerned about patient privacy.
  - Some patients walk past the two front desks, directly to back rooms.
    - Staff don't have time to put away patient information.
    - Can interrupt other patients' conversations.
  - The current layout allows patients to overhear other patient conversations.
- Conversations are in person or over phone.

Current Condition

Flowchart for patient path through the office:

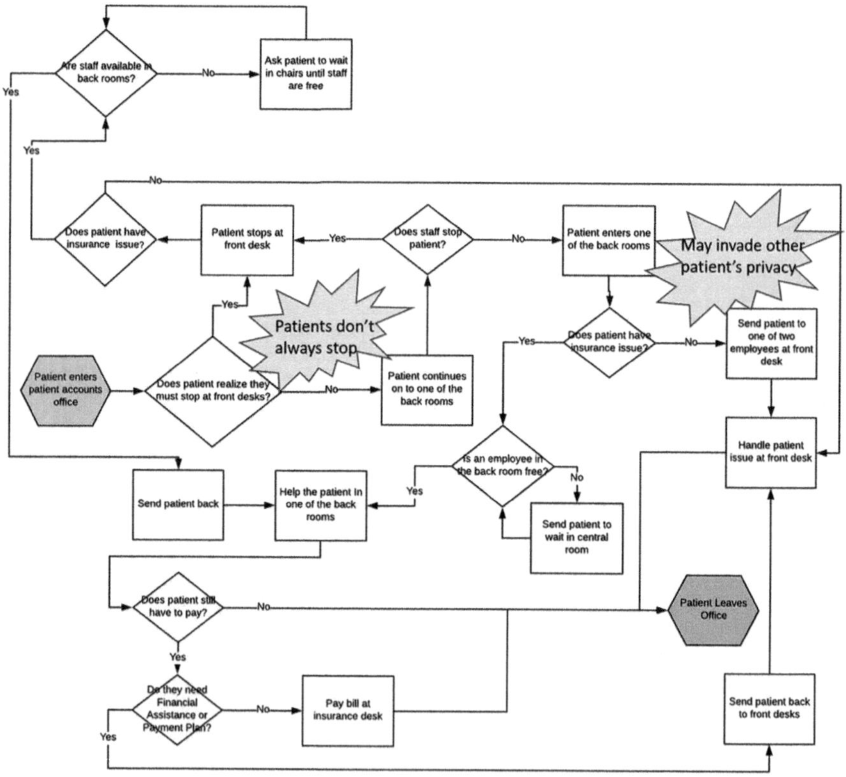

Heat Map of Noise Level with Conversation

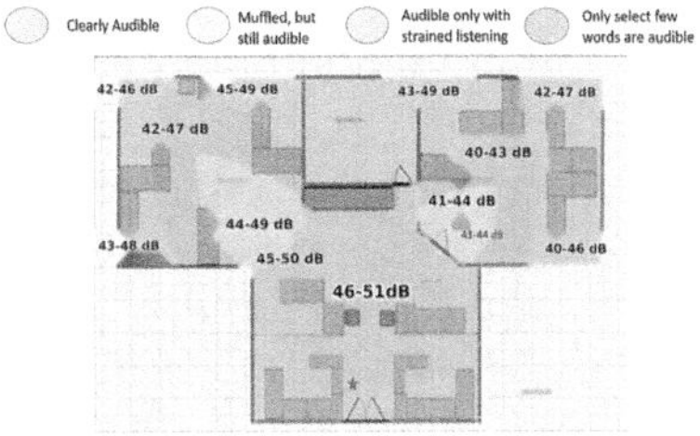

★Origin of conversation

Locations where conversation originating at the blue star are audible (can be heard and understood):

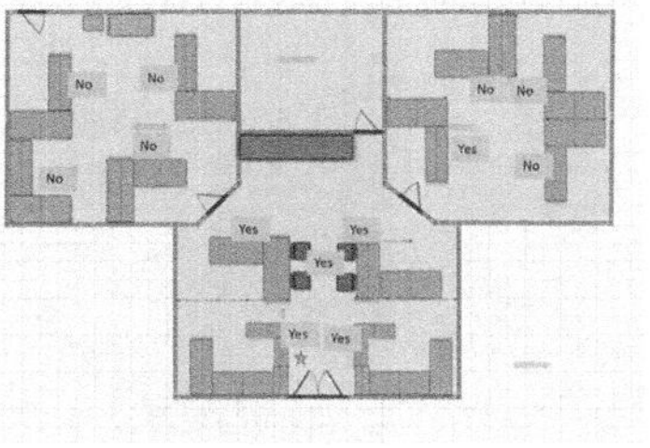

Note: For this test, a sample recording of a customer service call on YouTube was used.

Patient view entering the office:

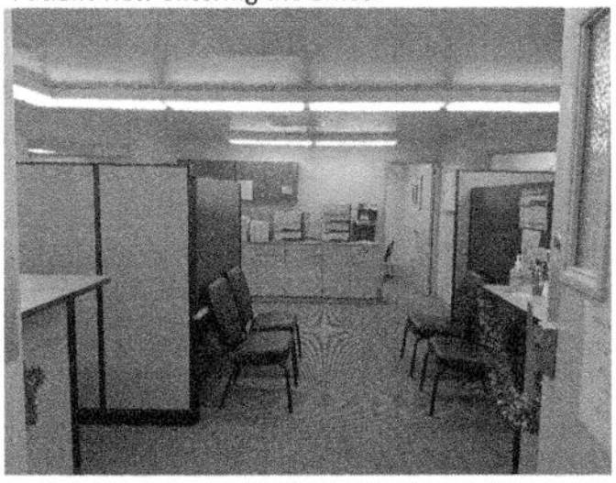

View of back left room:

SMART Goals:	• Make conversations inaudible in the patient waiting area. • Eliminate occurrences of patients not stopping at the front desk.

Root Cause Analysis

Patients Passing Front Desk:

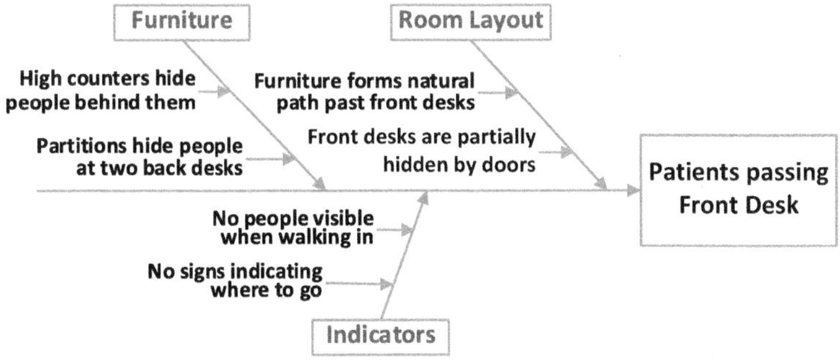

Patients Hearing Other Patient Conversations:

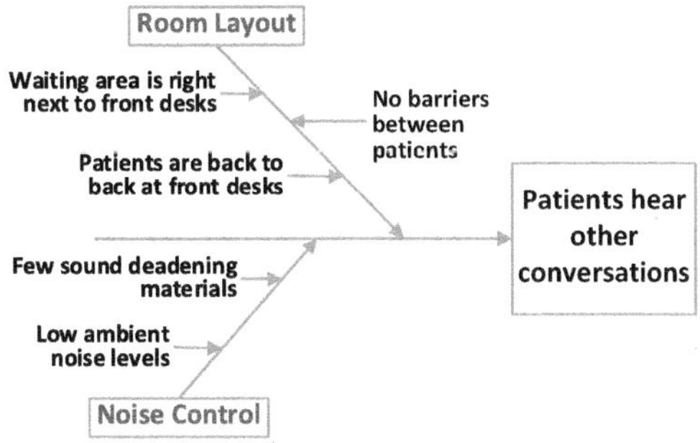

Proposed Countermeasures
Target Condition:

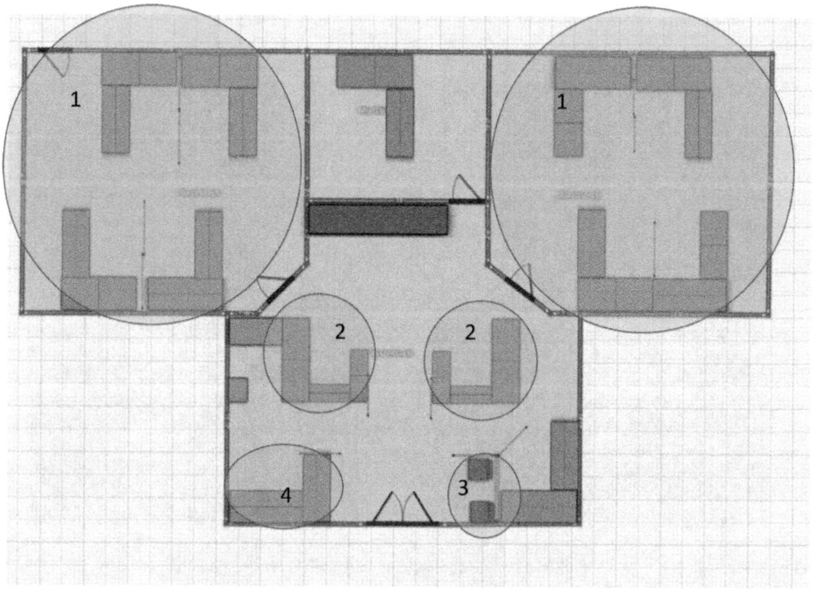

Back rooms rearranged with employees in the middle, patients on the outside.

- Will provide patients with more privacy by moving them to farther apart and more secluded locations.

Front Desks rearranged to face people entering the office.

- Provides a natural barrier to patients walking in.
- Makes employees visible to patients entering the office.

Waiting area is moved away from the center of the office.

- Moves idle patients farther away from where noise concentrates.

Third desk for potential new hire moved to the front of the room.

- Allows main front desks to be moved to the back of the room.
- Oriented to provide a third place to help patients if needed.

Noise Countermeasures:

- White noise machine near waiting area to mask conversational noise.
- Plants organized strategically to reduce conversational noise travel.
- Fans brought near patient locations to mask conversational noise.

Implementation Plan

Action	Responsible	Deadline	Expected Outcome	Actual Outcome
Get office approval for the new layout.	Ana	Monday (week 2)	Office approves the new layout.	Expected outcome met.
Get engineering approval on the new layout.	Ana	Monday (week 2)	Engineering approves the new layout.	Expected outcome met.
Move desks in back rooms into new position.	Ana, office staff, engineering	Tuesday (week 2)	Desks are located as denoted in the target state.	Expected outcome met.
Implement partitions in back rooms.	Ana, office staff, engineering	Tuesday (week 2)	Partitions implemented as denoted in the target state.	Expected outcome met.
Move desks in the main office.	Ana, office staff, engineering	Tuesday (week 2)	Desks located in the main office as denoted in the target state.	Expected outcome met.
Implement a new desk in the main office.	Ana, office staff, engineering	Tuesday (week 2)	A new desk is implemented as denoted in the target state.	Expected outcome met.
Implement partitions in the main office.	Ana, office staff, Engineering	Tuesday (week 2)	Partitions implemented as denoted in the target state.	Expected outcome met.
Implement a new waiting area.	Ana, office staff	Tuesday (week 2)	Waiting area implemented as denoted in the target state.	Expected outcome met.
Implement noise control countermeasures.	Ana, office staff	Tuesday–Thursday (week 2)	Noise countermeasures are in place.	Expected outcome met.
Track the number of patients passing the front desk.	Ana	Tuesday–Thursday (week 2)		Study completed.
Check the audible noise level at patient locations.	Ana, office staff	Thursday (week 2)		Study completed.

Effects Confirmation

Noise Target:

Locations where conversation originating at the grey star is audible (can be heard and understood), using the same recording as the current state:

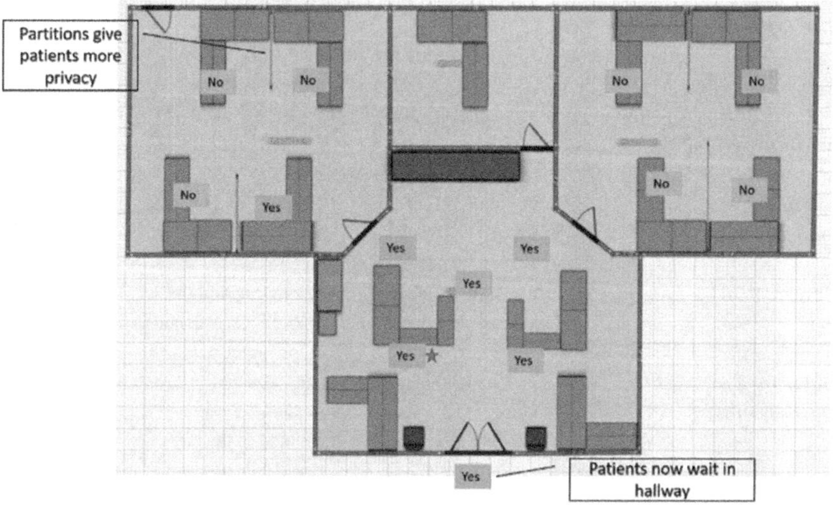

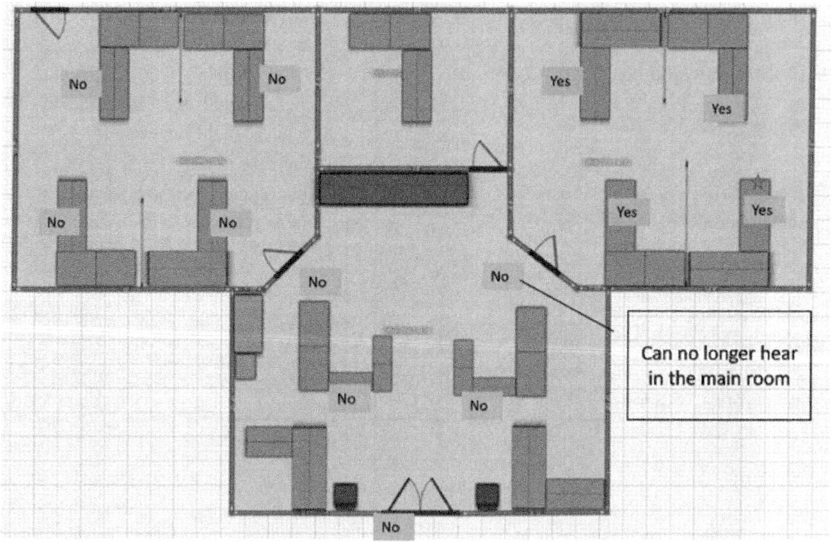

Results from the staff survey (scale of 0 = no change to 5 = much improved):

- Patient ability to know where to go: 4.7
- Patient privacy in back two rooms: 4.4
- Patient privacy in front room: 3.1
- Staff safety: 4.7
- Many patients commented on the more inviting feel and clarity of knowing where to go in the new layout.

Patient flow target: **Zero patients passed the front desks!**

Patient view entering the office:

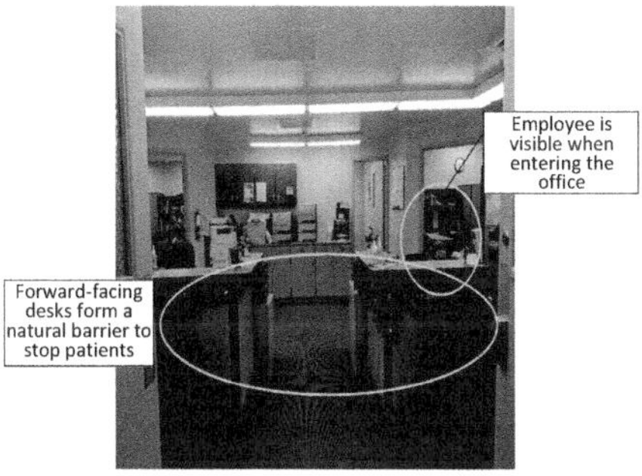

View of back left room:

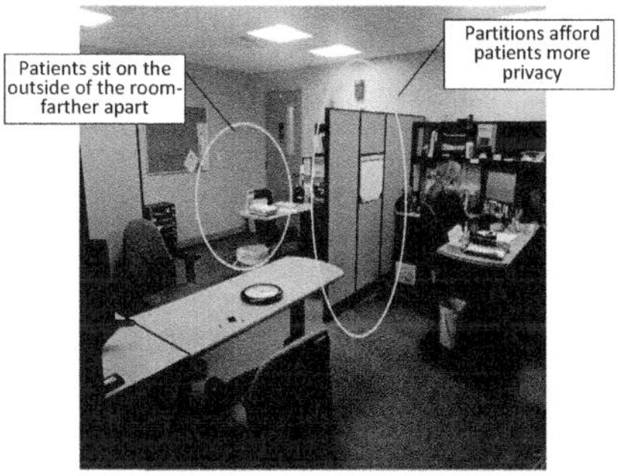

Follow-Up Actions

Action	Responsible Party	To Complete Action	Due
Continue to decorate office with noise control items: plants, artwork, fans, and white noise machines.	Office staff	Follow action.	In 2 weeks.
Determine how to orient the desk for the third Financial Counselor.	Office staff	Follow action.	In 2 weeks.
Keep track of patients passing the front desk.	Office staff	Follow action.	For 2 weeks.

Case Study 8.4: Business Office Revenue Cycle Improvement	Date: XX/XX/20XX
Sponsor: Clinic Business Office	**Author: Erika**

Background

- A clinic has three board-certified family medicine physicians, along with one physician's assistant, and two nurse practitioners.
- The clinic is open from 7:30 a.m. to 6:00 p.m., Monday–Friday.
- The clinic sees approx. 40 patients per day.
- The charge ticket is handled by seven people from the time it is generated to the time it is filed.
- The longer charge tickets go without being entered, the higher the risk of losing accuracy as well as reimbursement from insurance companies.

Current Condition

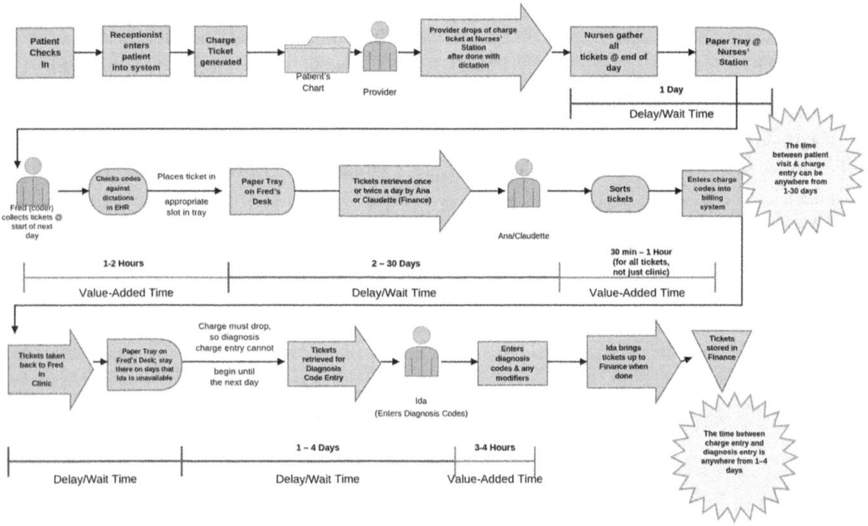

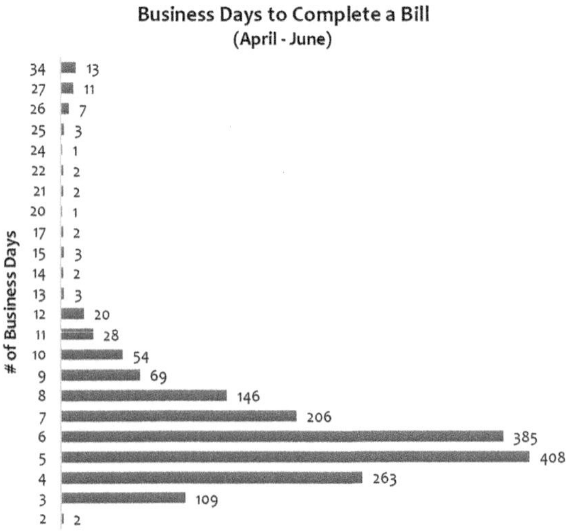

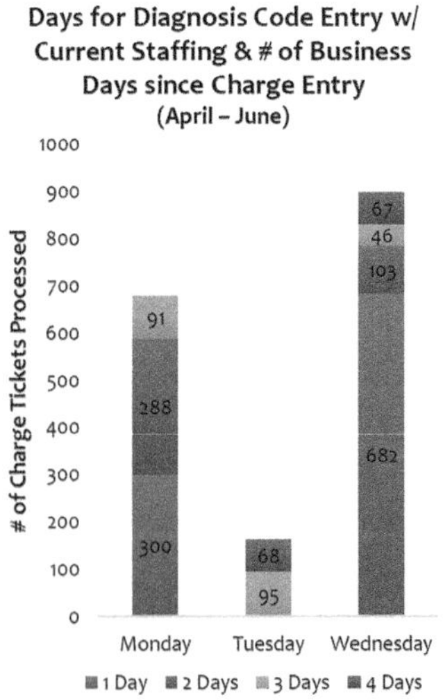

- It takes an average of 6.5 days to complete a bill, with approx. 4.7 days between patient visit and charge entry, and approx. 1.7 days between charge entry and diagnosis code entry (completing the billing cycle).
- Only 45% of bills from April to June were completed in 5 days or less.

SMART Goals:	• Decrease the number of bills completed in over 5 days by 50%. • Ensure that no ticket takes more than 1–2 days between charge entry and diagnosis code entry.

Root Cause Analysis

The clinic billing cycle is exceeding the desired time frame
- ➡ The charge tickets are not being processed quickly in the charge entry stage
  - ➡ Those doing the charge entry are spending more time organizing & sorting all the charge tickets (clinic, hospital, ER, swing), in order to more quickly & easily enter codes, than actually entering the codes
    - ➡ Few tickets come already organized or sorted
      - ➡ There is no standard in place for a sorting system among staff

The clinic billing cycle is exceeding the desired time frame
- ➡ The charge tickets are not being processed quickly in the charge entry stage
  - ➡ Tickets are not always retrieved by or brought to those doing charge entry
    - ➡ There is no standard system in place for the delivery/retrieval of the charge tickets

The clinic billing cycle is exceeding the desired time frame
- ➡ The charge tickets are not being processed quickly in the diagnosis code entry stage
  - ➡ The tickets sit on Diana's desk on days that there is no one available specifically for diagnosis code entry
    - ➡ Diana does not have the time to get to all the charge tickets waiting for diagnosis code entry & the person in charge of the clinic diagnosis entry only comes in 2 or 3 times a week

Proposed Countermeasures

Counter Measures	Needed Actions	Responsible	Expected Outcome
1. Create a standardized sorting system for Fred and the Health Information Management staff.	• Go over the best sorting method with Ana. • Go over new protocol w/staff. Put together folders or trays to place sorted documents into.	Erika Erika	Decrease the time spent on charge entry, as all charge tickets will already be sorted.
2. Have a system in place for the delivery and retrieval of charge tickets.	• Find the two or three best times in the day for Fred to deliver tickets or for Claudette/Ana to retrieve tickets. • Ensure that the new system is being followed.	Erika, Ana, Claudette, and Fred Erika, Ana, Claudette, and Fred	A standard system of delivery/retrieval will ensure that the tickets waiting for charge entry are taken from Fred's desk 2 or 3 times a day, making it possible for more charges to be entered.
3. Have the person entering diagnosis codes come in Monday, Wednesday, and Friday, or possibly train someone to cover Fridays or when Ida is gone.	• Discuss possible staffing options.	Ida, Danny, Grace, and Bill	Having one more day/week to enter diagnosis codes will ensure that stage of the process is completed in 1–2 days.

Target Condition:

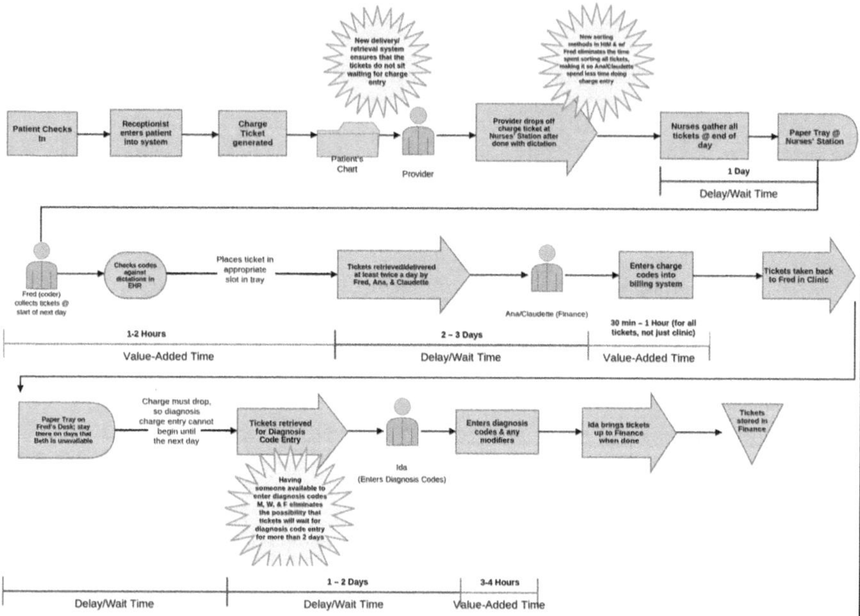

Implementation Plan

Action	Responsible	Deadline	Expected Outcome	Actual Outcome
Meet with Ana to create a sorting protocol for all charge tickets.	Erika and Ana	Tuesday (week 2)	Finalize protocol.	• Met with Ana and outlined all the ways to have the tickets sorted.
Make folder or tray that will make the new sorting process easier and self-explanatory.	Erika, Mary, and Fred	Tuesday (week 2)	Create a new sorting process.	• Added two trays in the Health Information Management office and labeled them for Inpatient, Outpatient, and Swing; will be sorted by date as well. • Fred will sort hospital by ER and Observations as well as by date; she will modify the trays already on her desk for this; will be sorted by date as well. • Ana now has a tray for facility charges and one for professional charges.

(Continued)

(Continued)

Action	Responsible	Deadline	Expected Outcome	Actual Outcome
Meet with Henri (DON) and discuss the ways of having the nurses fill out charge sheets in a more standard and easily understood way.	Erika and Henri	Tuesday (week 2)	Derive options to standardize charge sheets.	• Met with Henri and talked over possible options for making the charge sheets more standardized. • Decided that putting a grid on the back of the med/supply list would help to put the supply charge stickers in a more organized and standard order. • The grid is made and copied to the back of the list.
Talk with Fred, Ana, and Claudette about set delivery/retrieval times.	Erika	Tuesday (week 2)	Get buy-in on set delivery dates and times.	• Set so that Fred delivers tickets right before lunch (noon) and Ana/Claudette will go down and collect them at 1:30/2 p.m.

Effects Confirmation

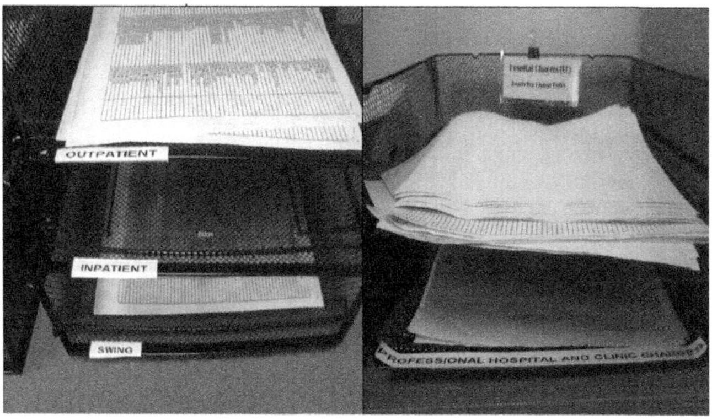

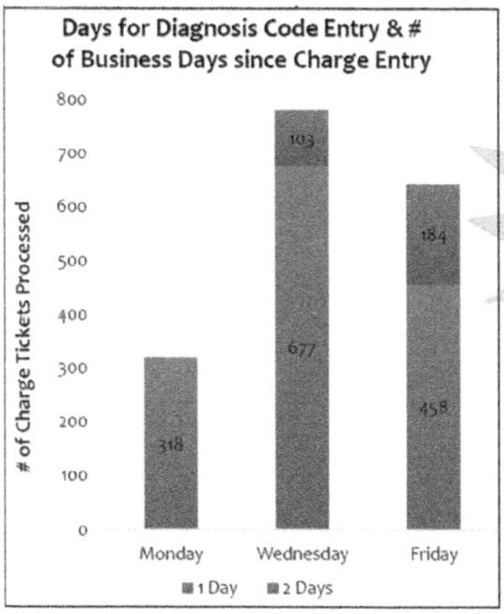

Changing the number of days spent entering diagnosis codes to 3 times a week, would cause 59% of the bills to be completed in 5 days or less instead of the earlier 45%

Follow-Up Actions

Action	Responsible Party	To Complete Action	Due
Continue to work with the staffing issue to ensure that someone is entering diagnosis codes M, W, F (ideally).	Bill, Danny, Grace, and Arthur	Look at possible cross-training opportunities as well as the possibility of Ida coming in 3 times/week.	In 3 weeks.
Monitor the queue going forward to ensure there are zero or very few tickets in the queue after diagnosis entry day and to see that the oldest tickets are completed first at all stages of the process.	Bill	Check with Fred and Ana/Claudette at the end of the day on M, W, F to see that this is being completed.	Ongoing.
Continue to remind Fred, Ana, and Claudette to retrieve/deliver tickets at least twice a day.	Claudette	Make sure the set times are working for everyone and find new ones if not. Fred delivers tickets before lunch, and Ana/Claudette retrieves tickets at 1:30/2:00.	Ongoing.

Case Study 8.5: Improve Patient Flow Between Clinic and Ancillary Services	Date: XX/XX/20XX
Sponsor: Ancillary services (Radiology and Lab)	Author: Arthur

Background

- Based on the outcome of clinic visit, providers sometimes order ancillary services.
  - Ancillary services: Radiology, Lab.
- Hospital registration is required when provider orders ancillary services.
  - Nurses provide stickers with service type and patient information to registration staff.
- It is not always communicated where patients are waiting for ancillary services.
- Time is wasted by service staff tracking down patients, or not realizing they are in waiting area.

Current Condition

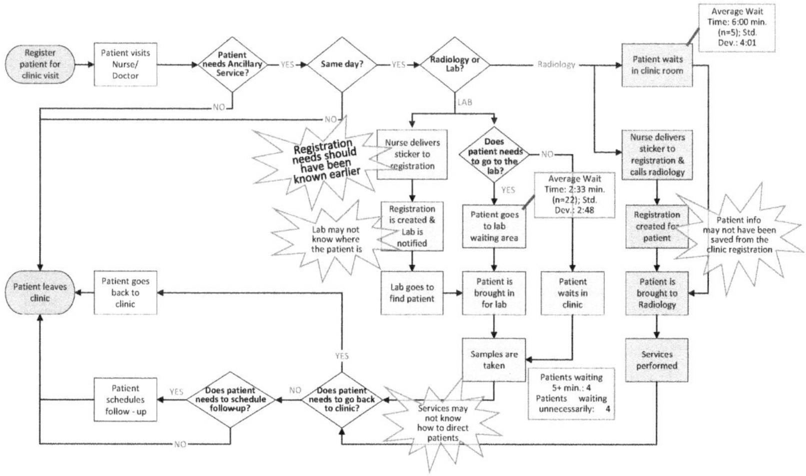

*Unnecessary waiting is defined as:

- Waiting for processes which could have been completed earlier.
- Waiting because staff are not aware that they are ready.

NOT:

- Waiting because services are occupied with other patients.

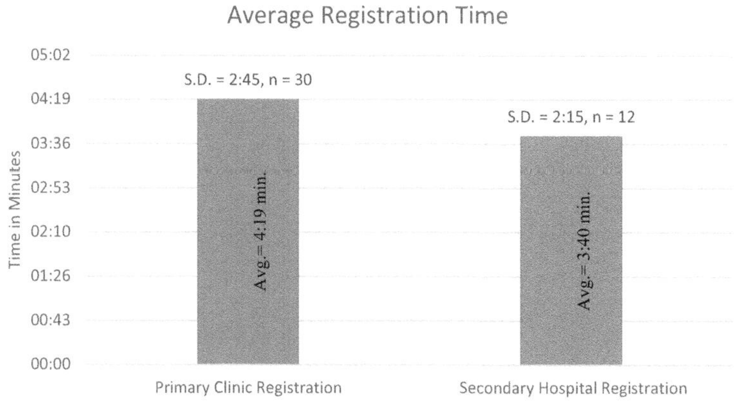

SMART Goals:	• **Completely eliminate unnecessary patient wait times for ancillary services.** • **Reduce patients' wait times for lab to less than 5 minutes.**

Root Cause Analysis

Patients are waiting in lab waiting area unnecessarily long because lab doesn't know they are there.

↳ **WHY→** Labs are told patients are "in the clinic" and don't know how to interpret this.

↳ **WHY→** Being told the patient is "in the clinic" doesn't actually provide their location – sometimes will be in the clinic, sometimes in the waiting area.

↳ **WHY→** Registration is making the call, but they don't actually know the patient location *(are guessing).*

↳ **WHY→** It is not always communicated to them if the lab has been notified of the patient location – assume it is their responsibility to let lab know.

Lab is guessing that patients are in waiting area and checking when they get a registration *(patients aren't always there).*

↳ **WHY→** Can't rely on other cues to tell them where the patient is in the system.

↳ **WHY→** Verbal cues don't always give an accurate patient location.

↳ **WHY→** Verbal cues aren't always coming from people who actually know the patient location.

↳ **WHY→** Nursing staff/provider doesn't always notify the lab of the patient location.

↳ **WHY→** Registration is assuming it is their job to notify the lab, and guessing that the patient is in the clinic.

Lab knows the patient will need a draw based on clinic visit type, but hospital registration isn't completed until after the clinic visit, leaving the patient waiting.

↳ **WHY→** Registration doesn't complete both registrations at once.

↳ **WHY→** They don't realize that the patient will need a lab.

↳ **WHY→** They don't know all visit types that automatically have a lab.

↳ **WHY→** They don't have access to information about what visits always have a lab; any they do know are learned through practice.

Proposed Countermeasures

Target Condition:

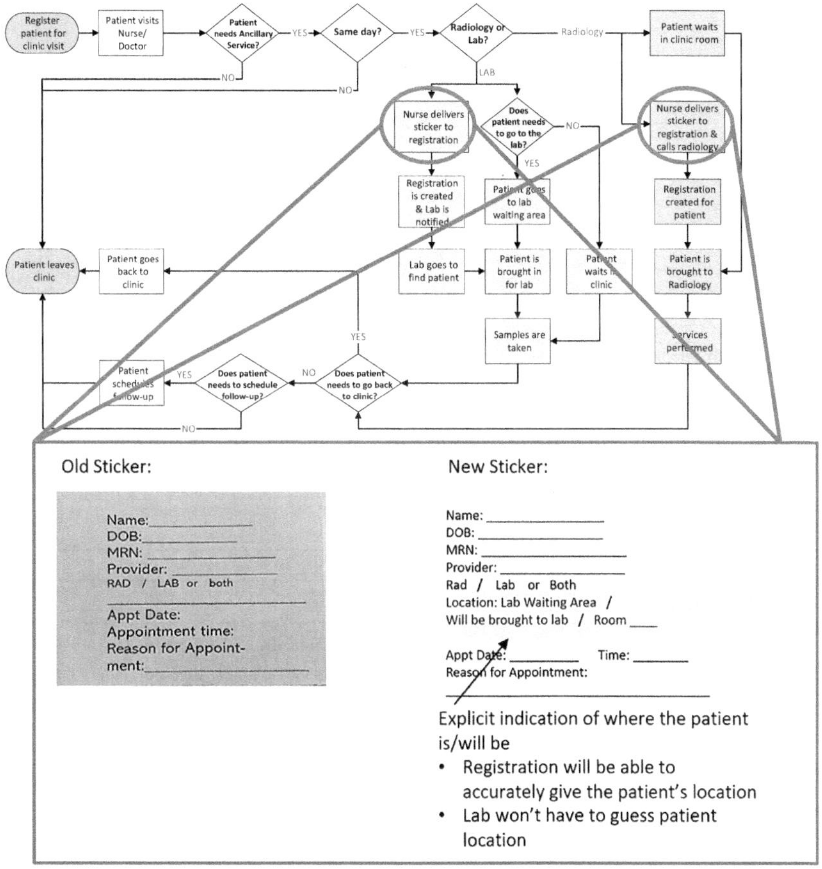

Implementation Plan

Action	Responsible	Deadline	Expected Outcome	Actual Outcome
Create stickers with added location identifier.	Arthur	Tuesday (week 2)	Stickers contain an identifier so registration staff can more accurately communicate the patient's location.	Expected outcome met.
Share new sticker updates with registration staff.	Arthur	Tuesday (week 2)	Registration staff understand what each location option indicates/what to communicate to lab.	Expected outcome met.
Provide new stickers to nurse staff.	Arthur	Tuesday (week 2)	Nurses understand what each option indicates and have stickers available.	Expected outcome met.
Share new sticker update with the ancillary staff.	Arthur	Tuesday–Wednesday (week 2)	Ancillary staff understand the new process and what changes to expect.	Expected outcome met.

Effects Confirmation

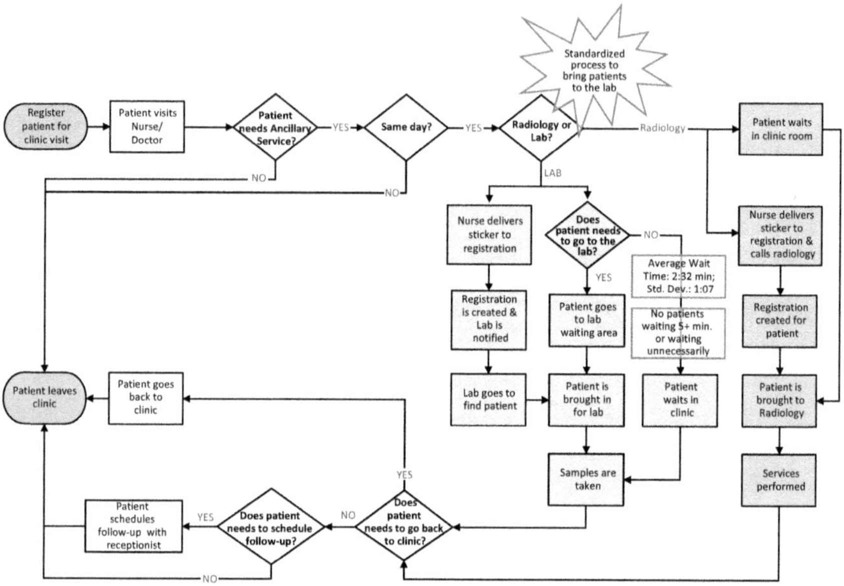

Follow-Up Actions

Action	Responsible Party	To Complete Action	Due
Continue to monitor and evaluate the impact of the countermeasures.	Bertha	Follow action.	For 1 month
Investigate the issue with mammograms printing incorrectly.	Dolly	Follow action.	In 6 weeks.
Evaluate the benefit of having a staff member dedicated to PT/Lab/Rad registration	Bertha	Follow action.	In 2 months.

9 Case Studies Related to Scheduling

A common source of ineffectiveness and inefficiencies in healthcare operations is timing and coordination. This chapter highlights five case studies centered on improving scheduling, coordination, and workflow synchronization across multiple hospital departments, including Pharmacy, Dietary Services, the Operating Room, and Clinics. Each RIE applied Lean and PDSA principles to streamline how people, tasks, and resources are scheduled, showing that better timing and coordination can significantly enhance both efficiency and patient experience.

Case 9.1 focused on pharmacy delivery schedules that created long gaps between medication restocks. The team found a 9-hour delay between morning and afternoon deliveries, leading to stockouts and interruptions in care. By redesigning delivery routes and adjusting print, pick, and delivery times based on medication demand, the team reduced average delivery time from 4.5 hours to 3 hours and doubled the number of daily deliveries with minimal added labor.

Case 9.2 examined staff scheduling in the Dietary Department, where inconsistent task assignments left 44% of staff time non-value-added. Through workload balancing, reassigning duties, and redefining an assistant role, the team increased time spent on required tasks from 56% to 70%, while non-value-added time dropped to 10%. The result was a more predictable and efficient schedule that better matched staffing levels to daily demand.

Case study 9.3 was previously presented and discussed in Chapter 2. This RIE tackled the challenge of reducing the number of late-start first-round surgery cases in the operating room department. Historically, 90% of first surgeries began late, creating ripple effects throughout the schedule. By carefully studying and timing the preparation activities of the perioperative nurses, anesthesiologists, surgeons, and operating room nurses, a synchronized morning plan for each role was piloted that increased on-time starts to 71% when most participants followed the new schedule.

Case 9.4 focused on meal delivery scheduling for residents in a long-term care facility. Mealtimes were inconsistent due to inefficient staffing and long travel distances. The team introduced a "hostess" role to support certified nursing assistants (CNAs), adjusted serving schedules, and proposed installing dishwashers in dining pods to reduce delays. The changes resulted in breakfast and lunch being served within 5 minutes of target times and improved staff–resident interactions.

The final case of this book deals with one of the most common patient complaints about our healthcare system: waiting. Patient waiting time is a sticky problem with many interacting factors that cause patients to feel they are waiting too long to see their provider. In this clinic, overbooking and poor alignment between providers and support staff led to long waits. By analyzing process flow and appointment patterns, the team recommended 30-minute minimum appointment blocks and daily planning

DOI: 10.1201/9781003670698-9

meetings between nurses and providers. Simulations showed potential to cut average wait times by 85%, from 16 minutes to just 3 minutes. The facilitator also provided several options for increasing clinic capacity to support the new scheduling approach to meet increasing patient demand.

These case studies show how important it is to assess the unit's capacity to handle the volume of work with not just the overall work demands, but *when* that work should or needs to happen, i.e., effective scheduling is the backbone of efficiency. They also highlight how much smoother operations can function when the timing and coordination of activities are systematically visualized and coordinated using data. These teams demonstrated that structured scheduling, grounded in A3 thinking and PDSA cycles, can reduce waste, improve communication, and align daily operations to better serve patients and staff. And, of course, no improvements can be realized without the collaboration of all the relevant parties, so communication, shared understanding, and agreement on the best course of action are vital ingredients to successful Rapid Improvement Events.

Case Study 9.1: Decrease Pharmacy Delivery Times	Date: XX/XX/20XX
Sponsor: Pharmacy	**Authors: Laura and Marco**

Background

- The hospital wants to optimize medicine use while assuring the safety of medication delivery.
- There have been 112 stockouts in the last 15 days.
- 348 items have gone critically low at least once in the last 15 days.

Current Condition

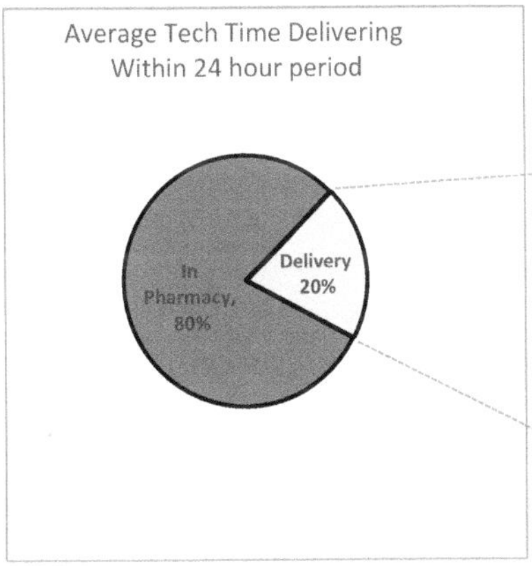

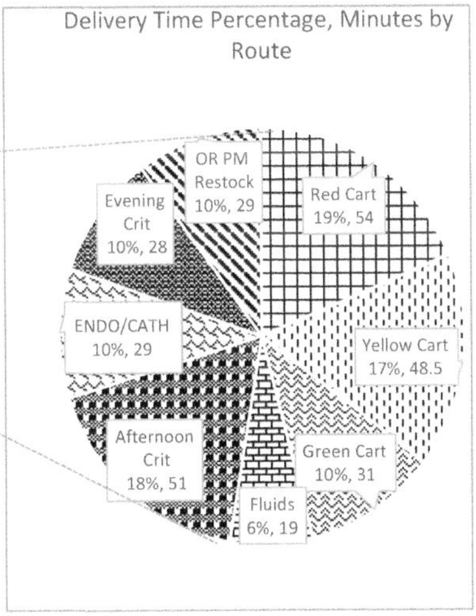

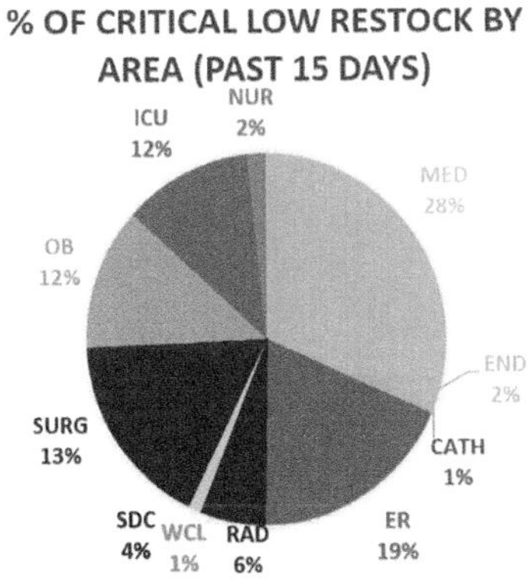

Below is a graphical depiction of the current timelines over a 24-hour period of the techs that deliver.

- X's and +'s represent the times each tech is in.
- + times represent the current times that techs are available for deliveries.

Time	0100	0200	0300	0400	0500	0600	0700	0800	0900	1000	1100	1200	1300	1400	1500	1600	1700	1800	1900	2000	2100	2200	2300	2400
Tech 1							X	+	+	+	+	+	+	+	+	+	+							
Central Tech																	+	+	+	+				
Tech 3	+	X	X	X	X	X													+	+	+	+	X	X

Below is a graphical depiction of the current timelines over a 24-hour period of the techs that deliver.

- X's and +'srepresent the times each tech is in.
- + times represent the current times that techs are available for deliveries.

Time	100	200	300	400	500	600	700	800	900	1000	1100	1200	1300	1400	1500	1600	1700	1800	1900	2000	2100	2200
Tech 1							X	+	+	+	+	+	+	+	+	+	+					
Central Tech																	+	+	+	+		
Tech 3	+	X	X	X	X	X													+	+	+	+

Demand By Route

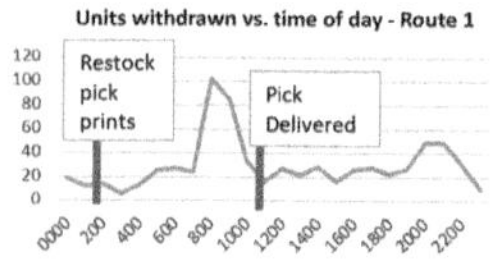

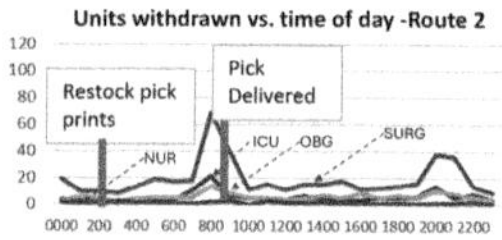

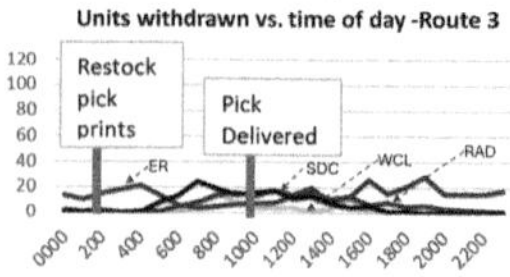

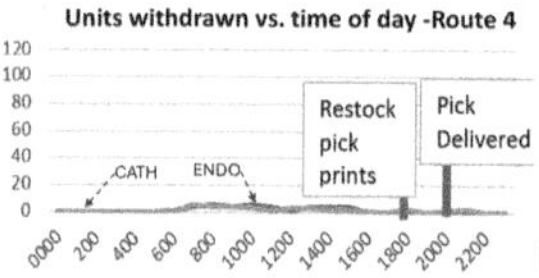

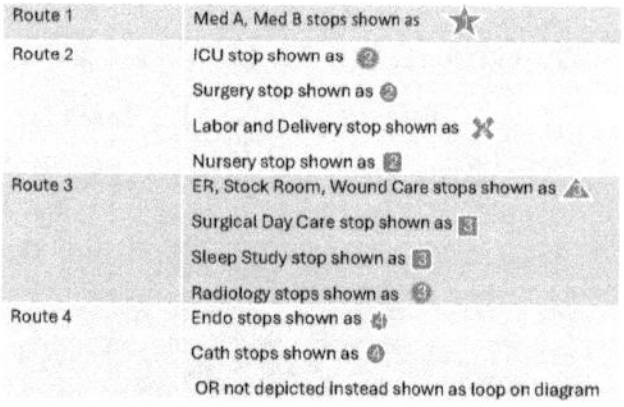
Route 1 Med A, Med B stops shown as
Route 2 ICU stop shown as
Surgery stop shown as
Labor and Delivery stop shown as
Nursery stop shown as
Route 3 ER, Stock Room, Wound Care stops shown as
Surgical Day Care stop shown as
Sleep Study stop shown as
Radiology stops shown as
Route 4 Endo stops shown as
Cath stops shown as
OR not depicted instead shown as loop on diagram

Legend for the following routing maps:

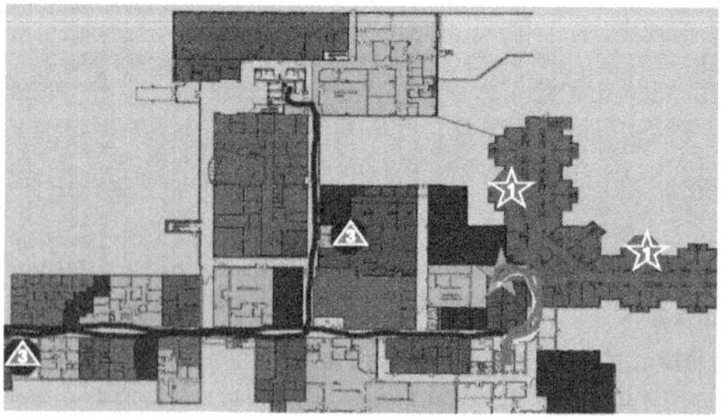

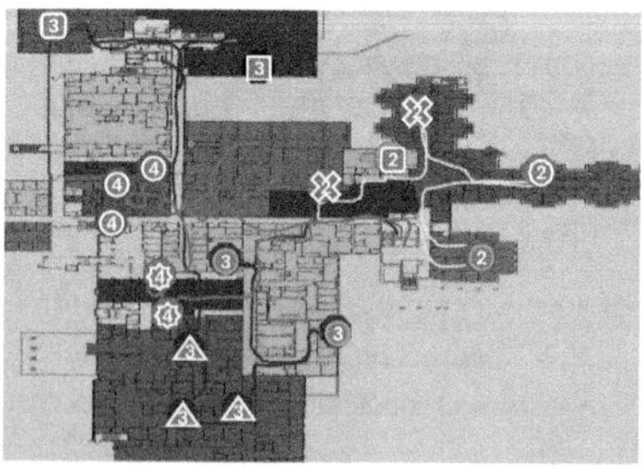

Route 1	**Med A, Med B stops shown as**
Route 2	**ICU stop shown as**
	Surgery stop shown as
	Labor and Delivery stop shown as
	Nursery stop shown as
Route 3	**ER, Stock Room, Wound Care stops shown as**
	Surgical Day Care stop shown as
	Sleep Study stop shown as
	Radiology stops shown as
Route 4	**Endo stops shown as**
	Cath stops shown as
	OR not depicted instead shown as loop on diagram

Restock process map

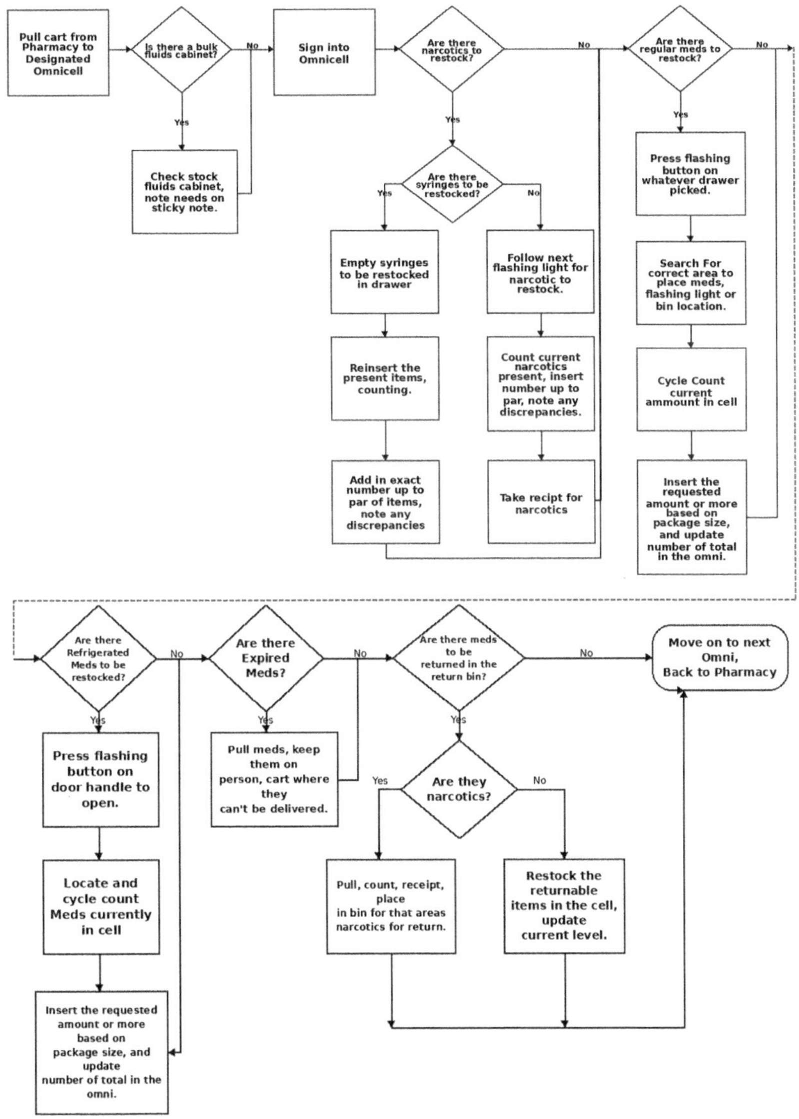

SMART Goal:	• Reduce the time between Restock Pick Printed to Pick Delivered from 4.5 hours to 3.5 hours.

Root Cause Analysis

Critical Low runs occur twice a day.

- Omnis run low on items during the day.
- 9–11 a.m. restock does not stock enough items to get through the day.
- 9 hours gap exists between pick print and pick delivery.

Proposed Countermeasures

Courses of Action:

1. Set print, pick, delivery times, and delivery routes based on daily medication demand patterns.
2. Set print and pick time to between 5 and 7 a.m. while maintaining current delivery routes and times.
3. Set print, pick, and deliver times based on daily medication demand patterns and maintain current delivery routes.

Courses of Action	Pros	Cons
1	• Largest potential for impact. • Fits in the current schedule of techs and pharmacists. • Largest potential to reducing interruptions by nursing staff. • Opportunity for route standardization.	• Techs must learn new delivery routes. • A new list must be built for print times.
2	• Easiest to implement, and requires changing on restock schedule by 3–5 hours. • Increased amount of coverage captured by the current report.	• Lowest potential impact. • No expected reduction in delivery time.
3	• Easier to implement than option 1. • Doesn't require delivery techs to learn new routes. • Expected to reduce interruptions by nursing staff.	• Requires change to tech 1 schedule, potentially other tech schedules.

Recommendation: Course of Action 1, pilot test for 2 days

Course of Action 1

- Set print, pick, delivery times, and delivery routes based on daily medication demand patterns.
  - Extracts maximum capture of meds usage while decreasing interruption of nurses.
  - Does not substantially increase workload for any tech.

For the new routes, the following adjustments are made for delivery times:

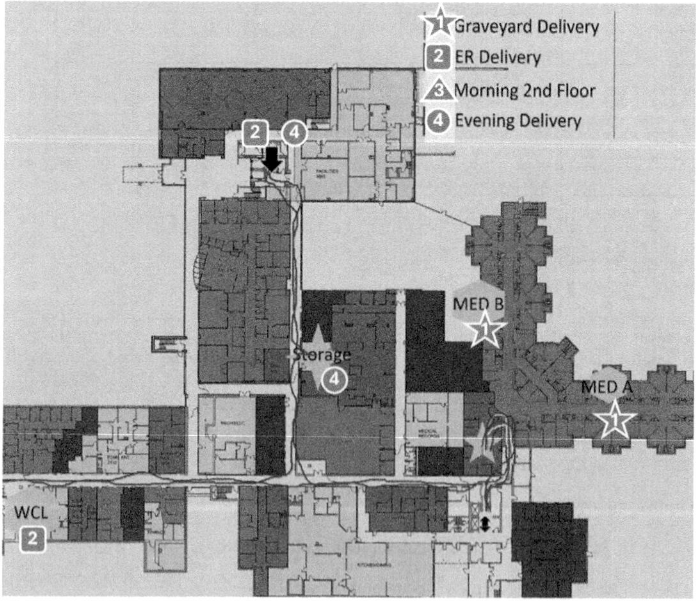

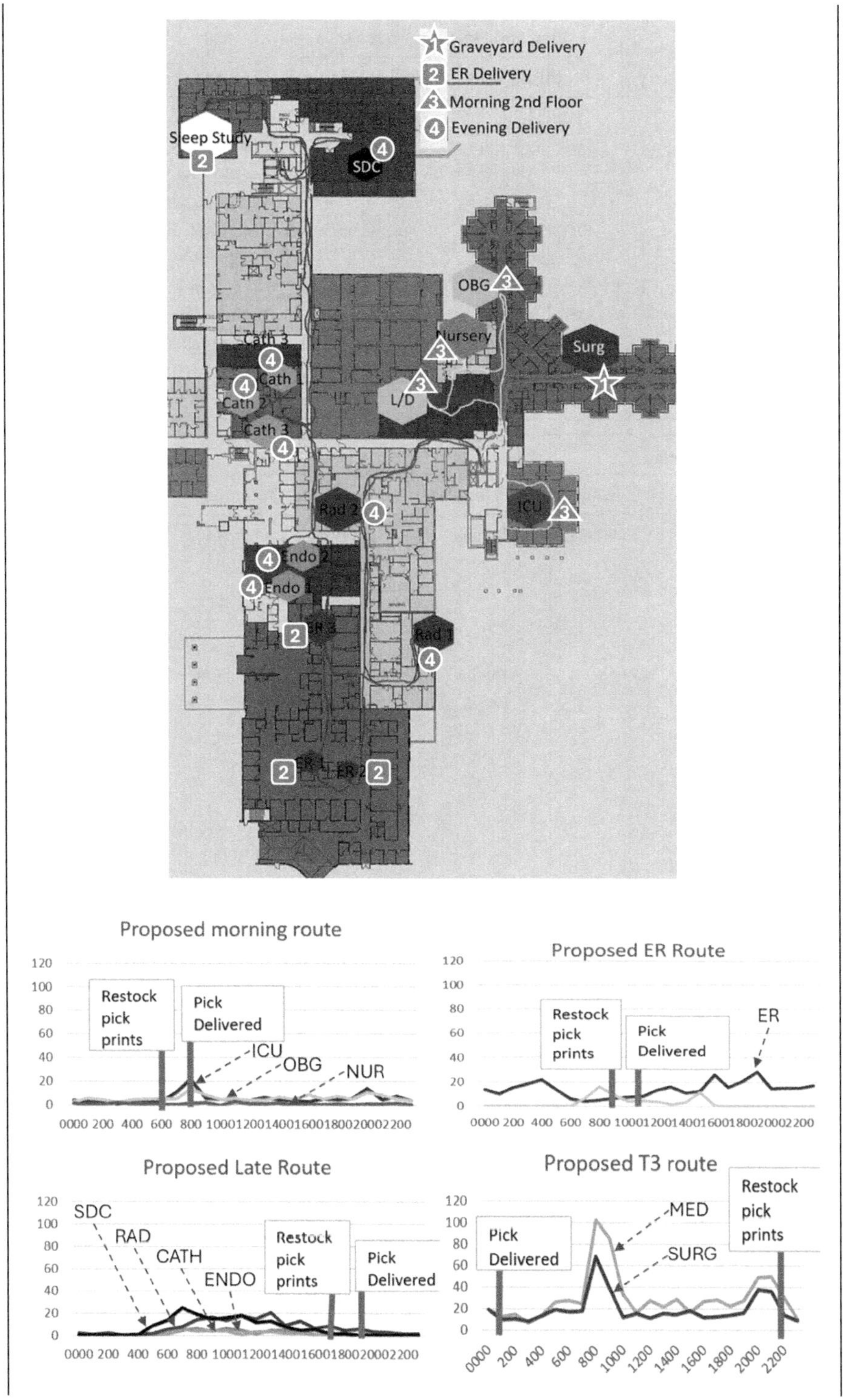
Graveyard Delivery
ER Delivery
Morning 2nd Floor
Evening Delivery
Sleep Study
SDC
OBG
Nursery
Surg
Cath 3
Cath 1
Cath 2
Cath 3
L/D
Rad 2
ICU
Endo 2
Endo 1
Rad 1
ER 1
ER 2
Proposed morning route
Restock pick prints
Pick Delivered
ICU
OBG
NUR
Proposed ER Route
Restock pick prints
Pick Delivered
ER
Proposed Late Route
SDC
RAD
CATH
ENDO
Restock pick prints
Pick Delivered
Proposed T3 route
Pick Delivered
MED
SURG
Restock pick prints

Route Name	Responsible Party	Pick and Delivery Start Times		Order of Omnicells in Route
		Proposed Time	Available Window	
Morning Route	Tech 1	Pick: 7 a.m. Deliver: 8 a.m.	7–8 a.m.	ICU, OBG (Above Med B), Nursery, L&D (behind desk of people who wear purple scrubs) – Route 2 w/o surgery
ER Route	Tech 1	Pick: 9 a.m. Deliver: 10 a.m.	7 a.m.–2 p.m.	ER 2, ER 1, ER 3, Wound Care, Sleep Study
Late Route	Central Tech	Pick: 5 p.m. Deliver: 6 p.m.	5 p.m.–3 a.m.	Rad 2, Rad 1, Endo 1, Endo 2, Cath 1, Cath 2, Cath 3, Same Day, Stock Room
Graveyard Route	Tech 3	Pick: 10 p.m. Deliver: 1 a.m.	10 p.m.–4 a.m.	Surgery, Med B, Med A

Implementation Plan

Action	Responsible	Deadline	Expected Outcome	Actual Outcome
Prepare documents for meeting Tuesday, including contingency plan and visual aid	Laura/Marco	Monday (week 2)	Supply all documentation for evaluation, signing off.	Had documents; will bring more.
Prepare a list of changes for Omni Tech	Laura	Monday (week 2)	Ensure smooth transition to and from the experimental state.	Resulted in 24 hour delay due to vacation, and complexity of changes. Deadlines have been updated in the following to reflect this.
Orient Techs and Pharmacists on changes to be made	Laura/Marco	Monday (week 2) -1 p.m.	Use visual aid during huddle to gain support of techs and pharmacists through socialization of the plan, and support during execution.	Techs were presented to; gave input afterwards. Expect more inputs to be made as time progresses.
Change print schedule	Omar	Tuesday (week 2) -4 p.m.	Transition to the experimental state with minimal hiccups; enable a new evening route to be printed and delivered in a previously determined window.	Finished at 2 p.m. Wed. CL at 2,200 kept.

(Continued)

(*Continued*)

Action	Responsible	Deadline	Expected Outcome	Actual Outcome
Gather Data	Marco, Laura	Tuesday (week 2) -4 p.m. to Friday (week 2) -12 p.m.	Have substantial evidence to determine if the current direction is improving the system state, and where we have issues.	Data collected and analyzed in spreadsheet.
Collect staff opinions, recommendations during huddles	Marco, Laura	1 p.m. TWRF	Adjust the system to better fit the current staff; ensure the conditions we've implemented are able to be accomplished effectively.	Routes adjusted based on input from techs.
Change print schedules back to original	Nana	Friday (week 2) – 7 p.m.	Transition back to the original state to ensure patient safety while system evaluation occurs.	Pushed back; extending trial period.

Effects Confirmation

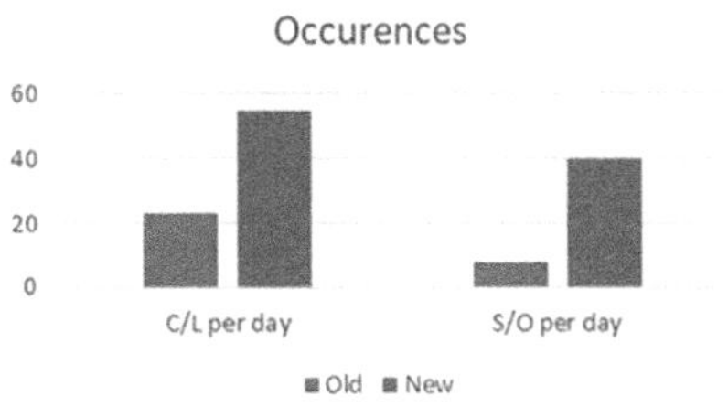

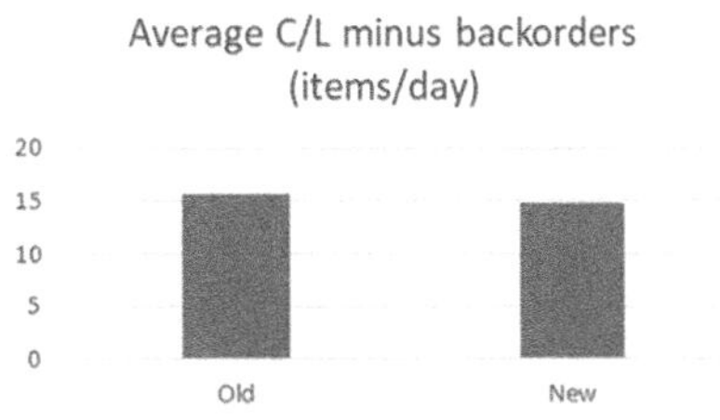

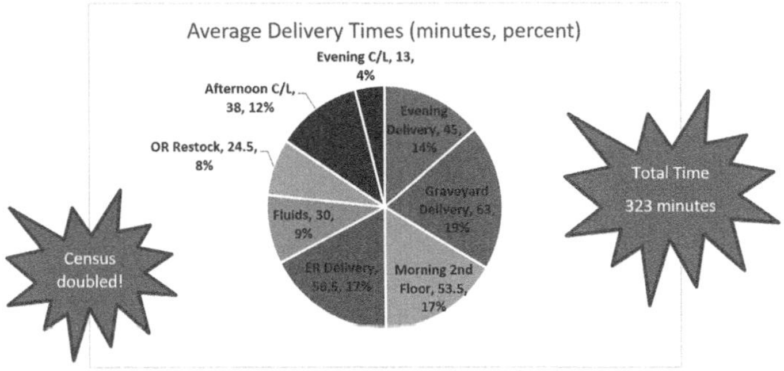

- Reduced time between Restock Pick Prints and Pick Delivery from 2–6 hours (avg. = 4.5 hours) to 2–3.5 hours (avg. = 3 hours).
- Doubled number of deliveries (more frequent deliveries) with only 33.5 additional minutes.

Follow-Up Actions

Action	Responsible Party	To Complete Action	Due
Continue logging restock delivery times.	T1, T3	Fill out the provided time sheets.	Ongoing.
Continue logging critical low delivery time.	T1, T3	Fill out the provided time sheets.	Ongoing.
Log census with delivery times.	T1, T3	Fill out the provided time sheets.	Ongoing.
Gather critical low-volume data over period.	Nana	Print Omni report of critical low.	Ongoing.
Analyze data.	Management/ engineering	Compare to previous route times and volumes.	In 3 months.

Case Study 9.2: Improve the Utilization of the Staff in the Dietary Department	**Date: XX/XX/20XX**
Sponsor: Dietary Department	**Authors: Andrea and Barry**

Background

- First Medical Center consists of the hospital, nursing home, retirement home, an outpatient clinic, and dental offices.
- The Dietary Department at First Medical Center provides breakfast, lunch, and supper to patients, nursing home residents, congregates from the retirement home, and medical center staff.
- The types of services performed by the dietary dept. are as follows: cart service to all non-ambulatory patients, tray service to the nursing home residents in the dining room, and cafeteria service to the retirement home congregates and medical center staff.
- There are four job categories the staff is divided into: cook, beverage server, dietary aide, and salad bar prep/baker.
- Scheduling problems create poor utilization of staff and tasks are not equally divided among employee job types.

Current Condition

Current Schedule by Job Type:

CURRENT SCHEDULE BY JOB TYPE:

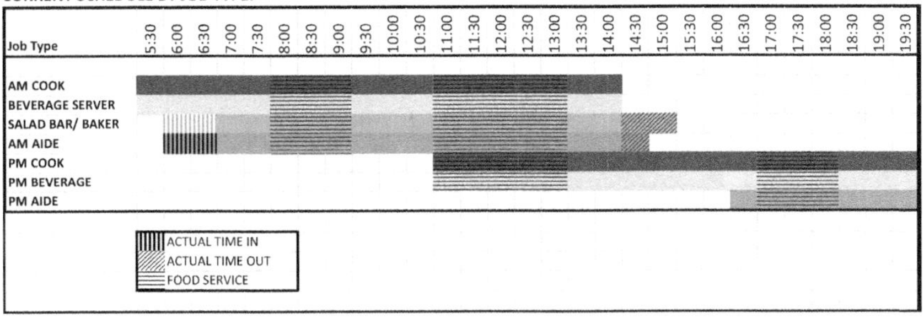

Major Problem:

- The amount of non-value-added time, gratuitous and neutral, is in excess of 44% of the total amount of time available in a day.

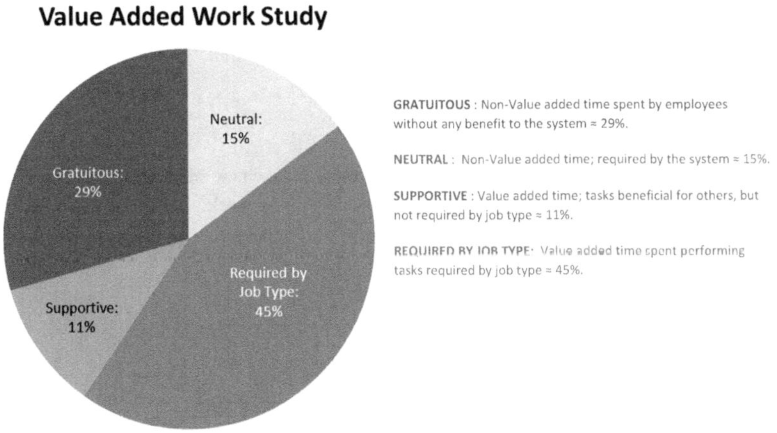

SMART Goal:	• Decrease the amount of gratuitous, non-value-added time as a percentage of total time to less than 15%.

Root Cause Analysis

Problem: **Employees wasting time**
Why? ↓
There isn't enough work to do.
Why? ↓
The morning staff does more work than their share.

Why? ↓	**Why? ↓**
The tasks are not evenly divided.	The morning staff arrives too early.
Why?↓	**Why? ↓**
There is no set schedule of tasks for the staff.	They are still using the schedule from the old facility.

Proposed Countermeasures

Counter Measures	Needed Actions	Responsible	Due	Expected Outcome
1. **Generate a new overall staff schedule.**	• Meet required hours for full-time employees.	Chantal, Barry, Andrea	6/7	Reduce the amount of non-value-added time in the schedule.
	• Create a new assistant position.	Chantal, Judy	6/8	
2. **Create schedule of tasks for each job type.**	• Schedule in breaks and adequate staff.	Andrea and Barry	6/11	Even the work load/ eliminate wasted time.
	• Meet with staff.	Chantal, Andrea, Barry	6/11	

Implementation Plan

Action	Responsible	Deadline	Expected Outcome	Actual Outcome
Meet with Chantal and Judy to get approval.	Andrea and Barry	Friday (week 1)	Get approval of the schedules.	Scheduled approved.
Meeting with staff about new schedules.	Andrea and Barry	Monday (week 2)	Train staff in new schedules.	Received feedback. Schedules revised.
Begin collecting data with new schedules in place.	Andrea and Barry	Tuesday (week 2)		Data collected.
Obtain feedback from staff.	Entire staff, Andrea, Barry	Tuesday (week 2)	Obtain feedback from staff.	Feedback obtained.
Make necessary changes and continue data collection.	Andrea and Barry	Wednesday (week 2)		New changes implemented.
Make new employee two-week schedules.	Chantal, Andrea, Barry	Thursday (week 2)	Make new employee two-week schedules.	Schedules created and approved.
Data collection and analysis.	Andrea and Barry	Thursday (week 2)		Data collected.
Present findings to staff and hospital management.	Andrea and Barry	Friday (week 2)		

Cost: *No expenditures required.*

Effects Confirmation

New overall **Staff Schedule**:

General 2-week Schedule (*using fictional names*):

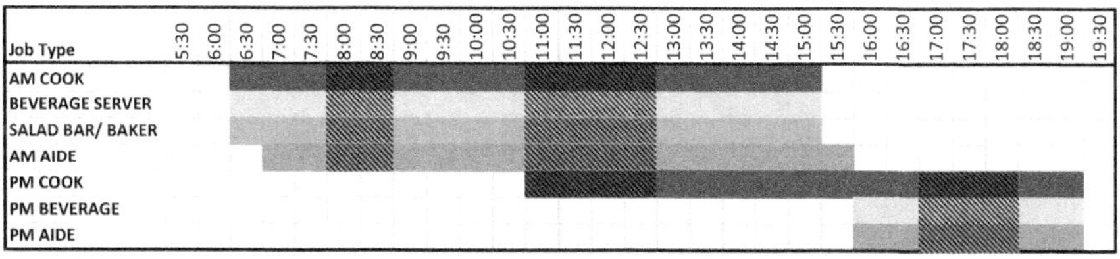

	SUNDAY	MONDAY	TUESDAY	WEDNESDAY	THURSDAY	FRIDAY	SATURDAY	SUNDAY	MONDAY	TUESDAY	WEDNESDAY	THURSDAY	FRIDAY	SATURDAY
AM COOK (6:30AM- 3PM)		KARLA	KARLA		KARLA	KARLA	KARLA	KARLA			KARLA	KARLA	KARLA	
AM COOK (6:30AM- 3PM)	JON			JON					JON	JON				JON
AM AIDE (7AM- 3:30PM)		CAROLINE	CAROLINE	CAROLINE		CAROLINE	CAROLINE	CAROLINE		CAROLINE	CAROLINE	CAROLINE		
AM AIDE (7AM- 3:30PM)	SAMMY				SAMMY				SAMMY				SAMMY	SAMMY
BEVERAGE (6:30AM- 3PM)			VICKY	VICKY	VICKY		VICKY	VICKY	VICKY	VICKY		VICKY	VICKY	
BEVERAGE (6:30AM- 3PM)						SAMMY					SAMMY			
BEVERAGE (6:30AM- 3PM)	LORI	LORI												LORI
SALAD/BAKER (6:30AM- 3PM)		CONNIE	CONNIE	CONNIE	CONNIE	CONNIE			CONNIE	CONNIE	CONNIE	CONNIE	CONNIE	
PM COOK (11AM- 7:30PM)	JENNIFER	JENNIFER	JENNIFER		JENNIFER		JENNIFER	JENNIFER	JENNIFER			JENNIFER		JENNIFER
PM COOK (11AM- 7:30PM)				SAMMY						SAMMY				
PM COOK (11AM- 7:30PM)						JON					JON		JON	
PM AIDE- (3PM-7PM)				LORI	LORI	LORI			LORI	LORI	LORI			
PM AIDE- (3PM-7PM)	ISABELLA													ISABELLA
PM AIDE- (3PM-7PM)		REBEKAH	REBEKAH				REBEKAH	REBEKAH				REBEKAH	REBEKAH	
Sharee's Assistant (11AM-3PM)				LORI	LORI	LORI			LORI	LORI	LORI			
PM BEVERAGE (4PM- 7PM)		ISABELLA			ISABELLA	ISABELLA			ISABELLA	ISABELLA				
PM BEVERAGE (4PM- 7PM)			BRENDAN	BRENDAN			BRENDAN	BRENDAN			BRENDAN	BRENDAN	BRENDAN	

New tasks by **Job Type**:

NEW AM COOK SCHEDULE		
6:30	7:00	BREAKFAST PREP/LUNCH REQUIRING EXTRA COOK TIME
7:00	7:30	
7:30	7:45	SETUP BREAKFAST SERVICE
7:45	8:00	PLATE FOOD FOR CARTS
8:00	8:30	PLATE & COOK FOR TRAYS TO RESIDENTS
8:30	8:45	BREAKFAST SERVICE
8:45	9:00	15 MIN-BREAK (AM AIDE COVER)
9:00	9:30	BREAK DOWN & CLEANUP
9:30	10:00	TODAY'S LUNCH PREP (NO DISHES)
10:00	10:30	
10:30	11:00	SETUP LUNCH & MEALS ON WHEELS
11:00	11:20	SERVE CONGREGATES
11:20	11:50	30 MIN-LUNCH BREAK (PM COOK COVER)
11:50	12:00	TRAYS FOR LUNCH
12:00	13:00	LUNCH SERVICE
13:00	13:45	BREAKDOWN & CLEANING
13:45	14:00	15 MIN-BREAK
14:00	14:30	PREP LUNCH FOR FOLLOWING DAY
14:30	15:00	

BAKING/CONNIE'S BREAKS		
6:30	8:30	SALAD BAR PREP
8:30	8:45	15 MIN-BREAK
8:45	9:30	SALAD BAR/SANDWICH BAR/ DESSERTS AND BAKING
9:30	10:00	
10:00	11:00	
11:00	11:30	BAKING AND CLEANING PROJECTS
11:30	12:00	
12:00	12:30	
12:30	1PM	1/2 HR LUNCH BREAK
13:00	13:45	FINISH BAKING AND CLEAN AREA
13:45	14:00	
14:00	14:30	15 MIN-BREAK
14:30	3PM	CLEANUP

NEW AM BEVERAGE SCHEDULE		
6:30	7:00	BREAKFAST BEVERAGES + COFFEE, ICED TEA & LEMONADE
7:00	7:30	(ALL BEVERAGES)
7:30	7:45	READY DINING AREA/BEVERAGE LINE (SUGAR, S&P, MILK, ETC.)
7:45	8:00	START SERVING BEVERAGES
8:00	8:30	SERVE BEVERAGES/TRAYS TO RESIDENTS
8:30	8:45	HELP COOK & RESIDENTS OR START DISHES
8:45	9:00	15 MIN-BREAK
9:00	9:30	CLEAN TABLES & FRONT OF HOUSE
9:30	10:00	LUNCH BEVERAGE SETUP
10:00	10:30	
10:30	10:45	MEALS ON WHEELS
10:45	11:00	CLEAN & STOCK FRONT BEVERAGES
11:00	11:20	TAKE TICKETS FROM CONGREGATES
11:20	11:50	TAKE 1/2 HOUR LUNCH BREAK
11:50	12:00	CLEAN FRONT OF HOUSE, PREPARE FOR RESIDENTS
12:00	12:30	BEVERAGE SERVICE & HELP W/ TRAYS
12:30	13:00	DISHES/ START CHARTING
13:00	13:45	CLEAN TABLES & FRONT OF HOUSE
13:45	14:00	15 MIN-BREAK
14:00	14:30	CLEAN SALAD/ SANDWICH BAR
14:30	15:00	GENERAL CLEANING AND SPECIAL PROJECTS

NEW AM AIDE SCHEDULE (7AM- 3:30PM)		
7:00	7:20	DISH ROOM SETUP & AM DISHES
7:20	7:40	10AM &2PM SUPPLEMENT TRAYS ONLY
7:40	7:50	TOAST
7:50	8:00	LOAD TRAYS ON CART
8:00	8:30	SERVE TRAYS TO RESIDENTS
8:30	8:45	15 MIN BREAK
8:45	9:00	COVER COOK FOR THEIR 15 MIN BREAK
9:00	9:30	DISHES & CLEANING
9:30	10:15	
10:15	11:00	PLATE DESSERTS & HELP SET UP SERVING AREA
11:00	11:20	HELP SERVE CONGREGATES TRAYS
11:20	11:50	STOCKING & DISHES/ HELP SERVICE LINE
11:50	12:00	DELIVER LUNCHES FROM CART
12:00	12:30	SERVE TRAYS TO RESIDENTS
12:30	13:00	1/2 HR LUNCH BREAK
13:00	13:30	DISHES & CLEANING
13:30	14:00	
14:00	14:15	15 MIN BREAK
14:15	15:15	EVENING SNACK/SUPPLEMENTS
15:15	15:30	LAUNDRY AND TAKE TRASH OUT

NEW SCHEDULE FOR PM COOK:		
11:00	11:20	CATCH UP DISHES
11:20	11:50	COVER AM COOK'S LUNCH
11:50	12:45	SERVICE LINE FIRST/CLEANING SECOND
12:45	13:00	15 MIN-BREAK
13:00	13:30	HELP WITH LUNCH CLEANUP
13:30	14:00	
14:00	14:45	COOKING; STOCKING; PREPPING;
14:45	15:00	15 MIN-BREAK
15:00	15:30	
15:30	16:00	SETUP FOR SUPPER SERVICE/CLEANING
16:00	16:30	AND BAKING PROJECTS
16:30	16:45	
16:45	17:00	LOAD CARTS FOR DELIVERY
17:00	17:30	SERVE SUPPER TO RESIDENTS
17:30	18:00	1/2 HR BREAK (WHEN FINISHED SERVING)
18:00	18:30	BREAK DOWN SUPPER & CLEANING
18:30	19:00	

NEW SCHEDULE FOR 4PM AIDES:		
16:00	16:15	CLEAR/CLEAN BEVERAGE CONTAINERS/COUNTERS
16:15	16:30	FILL OUT SUPPLY LIST
16:30	17:00	CATCH UP DISHES FROM COOK/BAKING/DESSERTS
17:00	17:45	HELP COOK SERVE TRAYS TO RESIDENTS
17:45	18:00	15 MIN-BREAK (FINISH SERVING FIRST)
18:00	19:00	CLEAN DISHES/ SWEEP & MOP ALL FLOORS

3PM BEVERAGE AIDE		
15:00	16:00	BAKE DESSERT FOR SUPPER
16:00	16:30	PLATE DESSERTS/READY TRAYS
16:30	17:00	PRE-POUR/READY BEVERAGES
17:00	17:45	SERVE BEVERAGES/TRAYS
17:45	18:00	15 MIN-BREAK (FINISH SERVING FIRST)
18:00	18:30	CHART/CLEAR/CLEAN TABLES
18:30	19:00	CLEAN & MOP FLOORS/ LAUNDRY & MOPHEAD

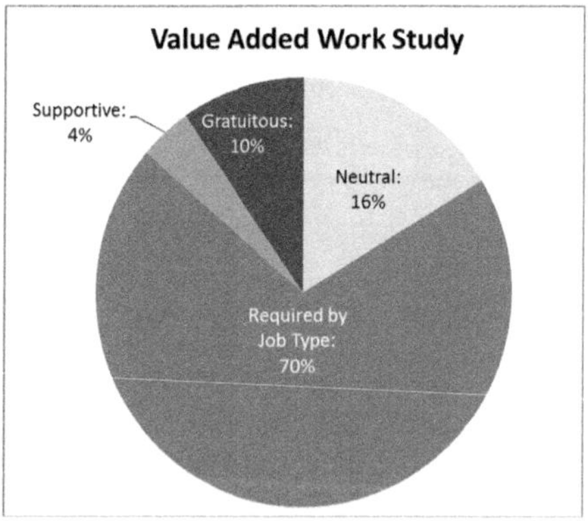

GRATUITOUS: Decreased to ≈ 10%.

NEUTRAL: Remained at ≈ 16%.

SUPPORTIVE: Decreased by moving to job specific ≈ 4%.

REQUIRED BY JOB TYPE: Increased to ≈ 70%.

Cost Analysis of Non-Value Added Time	# of Shifts Current	# of Shifts Proposed		
8 Hour Shift	6	5		
4 Hour Shift		1		
3 Hour Shift	1	1		
Total Hours Per Day	51	47		
% Time Wasted	29%	10%		
% Time on Breaks	15%	16%		
Amount of Time Wasted/ Day (hours)	14.8	4.7		
Amount of Time spent on Breaks/ Day (hours)	7.7	7.5	**Estimated Cost per Year:**	
Cost of Wasted Time (Estimated @ $14.00/hr average including benefits)	$207	$66	$10,767.12	$3,421.60
Cost of Time on Breaks/ day	$107	$105	$5,569.20	$5,474.56
Total Cost of Non-Value Added Time/ day	$314	$171	$16,336.32	$8,896.16

Follow-Up Actions

Action	Responsible Party	To Complete Action	Due
Use Erin as an assistant to Chantal.	Chantal and Erin	Create work tasks for Erin to complete 3 times a week for 4 hours as Chantal's assistant.	ASAP
Use the AM Aide or Beverage Aide to help serve food to stop congestion on the line.	Staff, Chantal, and Dorian	Work with cook to change the layout of the service line to allow room for someone to help plate and use Dorian to load carts.	In 1 week.
Make PM Cook shift 8 working hours.	PM Cooks and Chantal	Find tasks for the PM Cook to perform until 7:30.	In 1 week.
Use Aides to chart after every meal.	Chantal and Aides	Teach the same method for charting to all staff responsible for clearing tables after meals.	In 2 weeks.

Case Study 9.3: Late First-round Surgery Operating Room (OR) Arrival Rates	Date: XX/XX/20XX
Sponsor: Operating Room Department	**Authors: Arlene and Bret**

Background

- Patients arriving at the Operating Rooms (ORs) behind schedule cause later procedures to have compounding delays.
- First-round patients are expected to be in the OR by 7:15, but historical data shows that most arrive between 7:25 and 7:35 on average.
- The cost of running the OR is between $36 and $62 per minute.

Current Condition

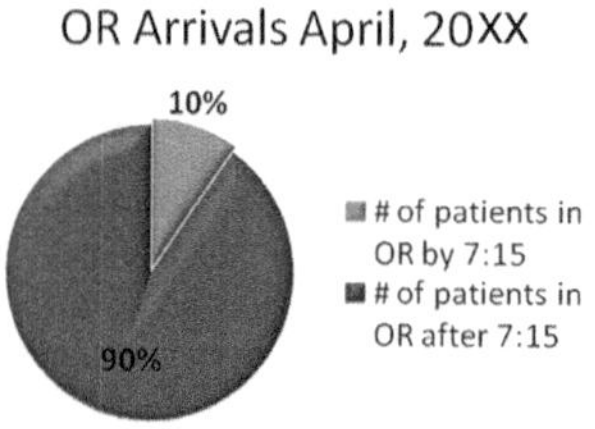

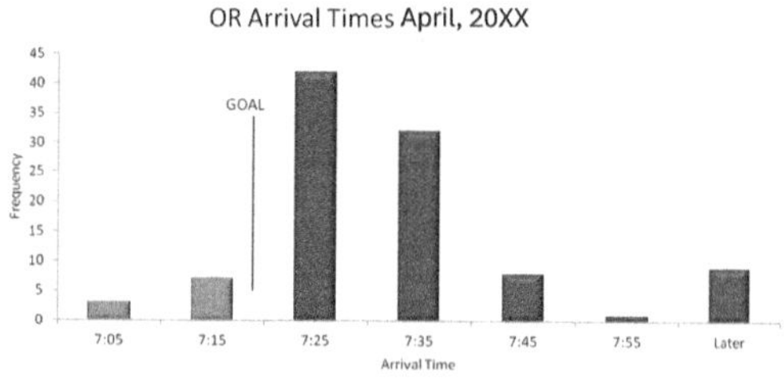

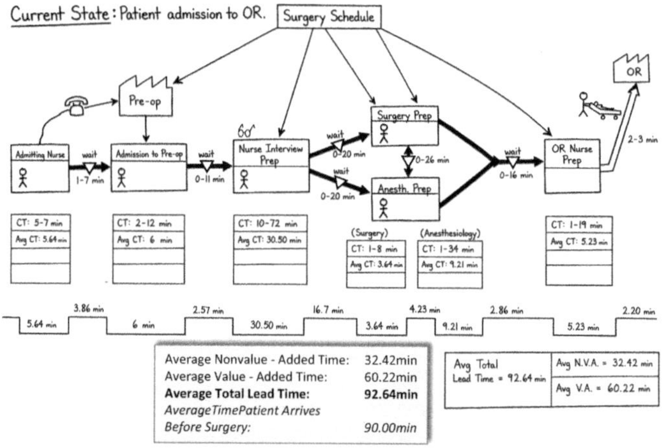

SMART Goal:	• **Reduce the average percentage of late first-round OR arrivals from 90% to less than 35%.**

Root Cause Analysis

Problem: Patients arriving late to the OR.

↳ **Why?** Anesthesiologist or Surgeon interview ends too close to 7:15 a.m.
  ↳ **Why?** The surgeon arrives at periop too close to 7:15.
    ↳ **Why?** The surgeon is unaware of the 7:15 goal.
      ↳ **Why?** The surgeon is unaware of what needs to happen after them.
        ↳ **Why?** There is no schedule designed for periop.

↳ **Why?** One Dr. waits for the other one to finish before coming into the room.
  ↳ **Why?** The surgeon and anesthesiologist arrived around the same time.
    ↳ **Why?** Insufficient communication/coordination
      ↳ **Why?** No established sequence.

- ↳ **Why?** Block **injection ended late.**
  - ↳ **Why?** Block injection (BI) had delayed start.
    - ↳ **Why?** The anesthesiologist arrived too close to 7:15 a.m.
      - ↳ **Why?** BI equipment unavailable.
        - ↳ **Why?** More than two BIs were scheduled at the same time.
- ↳ **Why?** Block injection took longer than usual.
  - ↳ **Why?** Intravenous (IV) bag complications.
    - ↳ **Why?** Patient concerns.
- ↳ **Why?** OR nurse prep ended after 7:15 a.m.
  - ↳ **Why?** The OR nurse left the OR too close to 7:15.
    - ↳ **Why?** The OR room was not ready by 7:15.
      - ↳ **Why?** No standard procedure.
        - ↳ **Why?** Staff can't find equipment.
          - ↳ **Why?** Not labeled or no one accountable for equipment locations.
- ↳ **Why?** OR periop interview is long.
  - ↳ **Why?** Patient questions.
    - ↳ **Why?** Missing patient info.
      - ↳ **Why?** Last minute doctor's orders.

Proposed Countermeasures

Target Condition:

Schedule: No Block Injection required

Time	Periop Nurse	Anesthesiologist	Surgeon	OR Nurse
5:30				
5:35				
5:40				
5:45				
5:50				
5:55				
6:00				
6:05				
6:10	Interview, IV prep, etc			
6:15				
6:20				
6:25				
6:30				
6:35				
6:40				
6:45		Check charts, interview patient		
6:50				
6:55			Check charts, interview patient	
7:00				
7:05				Check charts, interview patient, transport
7:10				
7:15				

Schedule: Block Injection required

Time	Periop Nurse	Anesthesiologist	Surgeon	OR Nurse
5:30				
5:35				
5:40				
5:45				
5:50				
5:55	Interview, IV prep, etc			
6:00				
6:05				
6:10				
6:15				
6:20				
6:25				
6:30		Check charts, interview patient, block injection		
6:35				
6:40				
6:45				
6:50				
6:55			Check charts, interview patient	
7:00				
7:05				Check charts, interview patient, transport
7:10				
7:15				

Implementation Plan

Action	Responsible	Deadline	Expected Outcome	Actual Outcome
1. Propose time schedules to OR and periop nurses.	Gert	Monday–Tuesday (week 2)	All nurses aware of the pilot plan.	Expected outcome met.
2. Propose time schedules to surgeons and anesthesiologists.	Gert	Monday–Tuesday (week 2)	All surgeons and anesth. aware of the pilot plan.	Expected outcome met.

(Continued)

(*Continued*)

Action	Responsible	Deadline	Expected Outcome	Actual Outcome
3. Answer any staff questions.	Arlene and Bret	Monday–Tuesday (week 2)	Questions have been answered.	Expected outcome met.
4. Post copies of new schedules.	Gert	Monday–Tuesday (week 2)	Schedules posted in several key OR and periop areas.	Expected outcome met.
5. Begin using proposed time schedules.	All nurses and doctors	Wednesday (week 2)	The pilot plan begins.	Expected outcome met.
6. Data collection of the pilot plan.	Bret and Arlene	Wednesday–Friday (week 2)	Data collected for all 7:30 surgeries.	Expected outcome met.
7. Present final report and follow-up actions.	Bret and Arlene	Friday (week 2)	The trial plan is either continued or not.	Trial took place.

Effects Confirmation

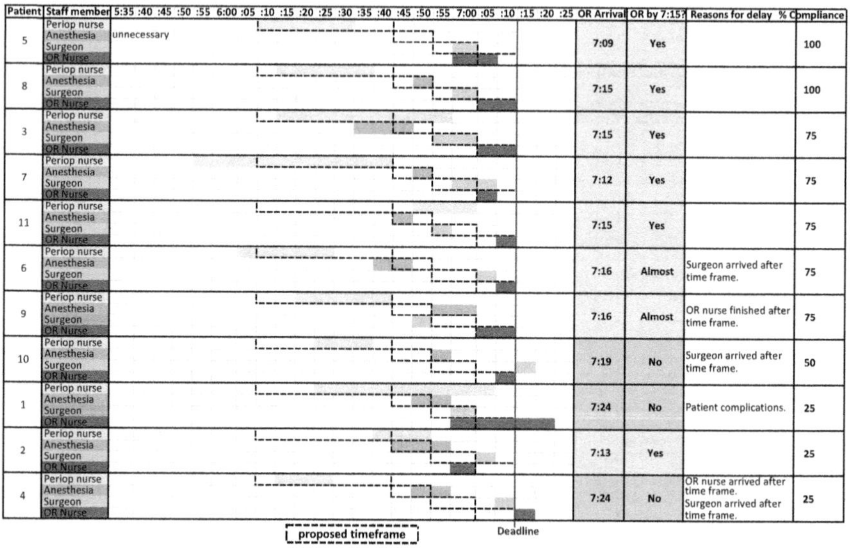

Results:

- Cases with ≥75% compliance (at least three out of the four stakeholders adhered to the schedule): **71% on-time.**
- Cases with <75% compliance (one or none of the four stakeholders adhered to the new schedule): **25% on-time.**

Follow-Up Actions

Action	Responsible Party	To Complete Action	Due
Continue the pilot study schedule for 1 or 2 weeks and confirm the pain block injection timeline. Then, assess the schedule and modify it as necessary.	Gert	Follow action.	In 2 weeks.
Continue to monitor compliance rates and give feedback to staff and doctors.	Gert	Follow action.	In 2 weeks.
Continue to work on getting buy-in from all staff members and doctors.	Gert	Follow action.	In 3 weeks.
Continue to identify additional sources of delay (e.g., lab tests, block injections, patient complications, etc.) and experiment with possible countermeasures.	Gert	Follow action.	In 4 weeks.

Case Study 9.4: Timely Delivery of Meals to Long-Term Care Facility	Date: XX/XX/20XX
Sponsor: Dietary Department	**Author: Earl**

Background

- The Dietary Department provides the following services: room service to all non-ambulatory patients in a 25 room hospital, cafeteria service to all health center staff, visitors, and the public, and food delivery and service to four pods of the long-term care facility.
- The long-term care facility requires CNAs to assist in the serving of meals and beverages to ≈70 residents.
- Due to the rapid growth of the community, all departments are understaffed.
- The layout of the facility requires excess travel and setup time to serve the residents, resulting in unreliable serving times for breakfast, lunch, and dinner.
- The current system for delivering the food and supplies and the cleaning and stocking of the four pods of the long-term care facility is inefficient and does not allow the dietary staff any additional time to assist the CNAs to guarantee the timely service of meals.

Current Condition

Flow chart of current process of serving breakfast to long-term care:

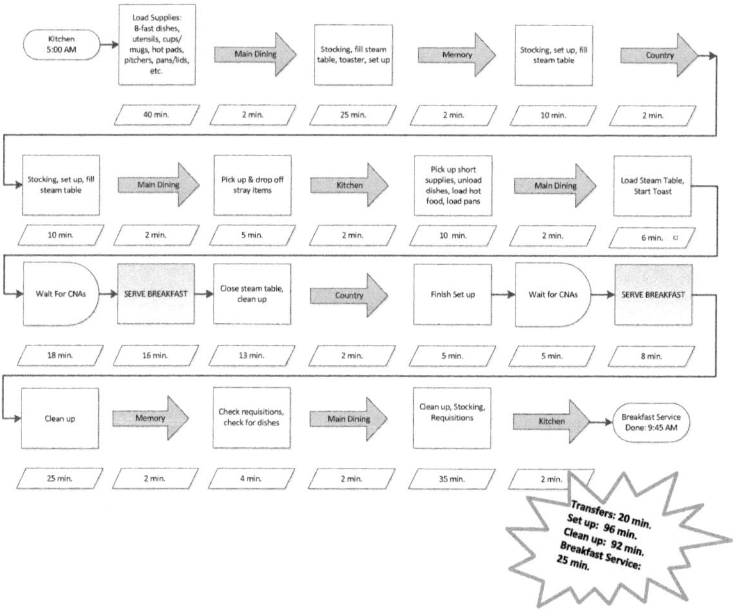

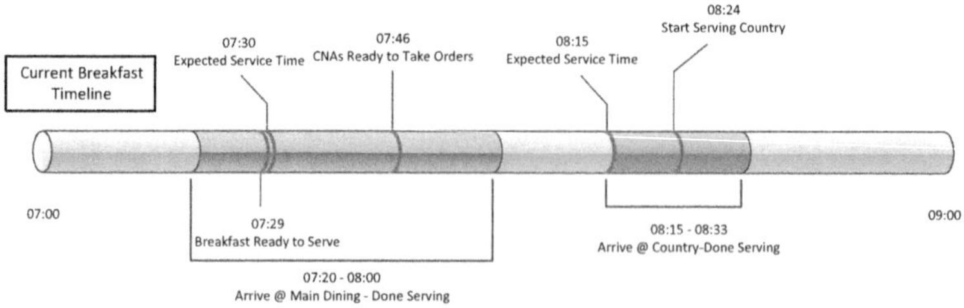

Major Problem: Due to the facility layout, staffing shortages and inefficient delivery of supplies, the long-term care facility is unable to provide meals at the specified times.

SMART Goal:	• **Increase the accuracy of meal service times to within 5 min. of the expected meal times.**

Root Cause Analysis

Problem: Food Service department cannot begin serving at the same times for each meal every day.

- **Why?** CNAs do not get beverages served and orders taken in time for meal service.
  - **Why?** They are busy getting residents ready or taking care of residents.
    - **Why?** Residents of the nursing home require different amounts of care daily.
      - **Why?** Cannot be changed, part of the system.

Problem: Food Service department cannot begin serving at the same times for each meal every day.

- **Why?** They are waiting for the CNAs to serve beverages and take orders.
  - **Why?** There is too much to accomplish to be ready in time to help the CNAs.
    - **Why?** The pods layout requires too much travel and setup to prepare for each meal.
      - **Why?** The setup takes place at the main kitchen and then is dispersed to the pods.
        - **Why?** The dishes and supplies are stored in the kitchen and must be delivered for each meal.

Proposed Countermeasures

Counter Measures	Needed Actions	Responsible	Expected Outcome
1. Get help in the main dining room to serve beverages and help #1s take orders.	• Cover for Alex's day off. • Talk to activities.	Bonnie or Colin Colin	Enable food service to serve breakfast on-time.
2. Use volunteers to assist CNAs in caring for residents.	• Contact volunteer programs to determine their availability. • Train volunteers to assist in serving beverages and working w/residents.	Danielle and management Colin and staff	Reduce the workload of CNAs and foodservice staff.
3. Provide locks on cabinets for sanitizing supplies.	• Contact maintenance to schedule install.	Danielle	Allow for Sanitizing Buckets to be stored in pods.
4. Bring all dishes to long-term care once daily.	• Talk to dishwashers to determine how/where to store. • Talk to 1s about bringing down all dishes. • Determine the number of dishes needed in each dining area.	Bonnie, supervisors, Earl Bonnie, supervisors Earl, staff, supervisors	Lessen the number of trips to/from kitchen and the need to sort dishes prior to each meal.
5. Leave sanitizing buckets in pods.	• Locks on cabinets. • Make sure buckets and supplies are available for each pod.	Danielle Supervisors	Fewer items to shuttle to kitchen; better cleaning habits.
6. Remove cereal and snack rack from the main dining room.	• Stock cereals in the main kitchen and devise a way to set out for CNAs.	Earl, staff	Fewer items to stock and move to/from the kitchen.

(*Continued*)

(*Continued*)

Counter Measures	Needed Actions	Responsible	Expected Outcome
7. Condiment caddies for pods.	• Order condiment caddies. • Fill Caddies.	Bonnie Earl, staff	Eliminate stock in pods and better display.
8. Determine the necessary stock of baked goods for breakfast.	• Establish how many muffins and Danish are being consumed.	Supervisors, staff	Eliminate waste and stocking; better quality.
9. Use #3s to clear everything at lunch from tables.	• Change employee job description and talk to staff.	Supervisors, staff	Slow down busing and give residents more time to finish; fewer employees needed to bus.
10. Bus and clean memory before country at lunch.	• Talk to #3s to change order of cleaning.	Supervisors, staff	Give residents a proper amount of time to eat.
11. Use #1s to serve beverages and take orders in country and memory and help in the main dining room.	• Talk to #1s and CNAs to change time they normally start serving; see the above proposed timeline.	Supervisors, Earl, Colin, and #1s	Get meals served in a timely manner and let CNAs focus on getting the residents to the meals.

Target Condition:

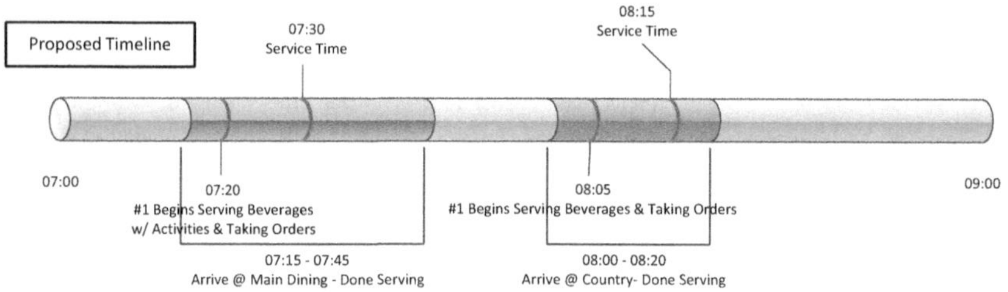

Implementation Plan

Action	Responsible	Deadline	Expected Outcome	Actual Outcome
Meet with Colin to discuss problems with the CNA time frame.	Earl and Colin	Wednesday (week 1)	Discuss problems with the CNA time frame to generate solutions.	Potential solutions generated.
Meeting with Bonnie and Fiona to discuss feasible ways to reduce setup times and free time to serve beverages to residents.	Earl, Bonnie, and Fiona	Thursday (week 1)	Reduce setup times and free time to serve beverages to residents.	Potential solutions generated.

(*Continued*)

(*Continued*)

Action	Responsible	Deadline	Expected Outcome	Actual Outcome
Meet with Danielle to discuss ideas for beverage service to residents.	Earl and Danielle	Thursday (week 1)	Generate ideas for beverage service to residents.	Potential solutions generated.
Provide a list of potential changes to food service staff.	Earl	Thursday (week 1)	Get buy-in from staff.	Staff decided on best ideas to implement.
Provide a preliminary implementation plan to management.	Earl	Monday (week 2)	Get buy-in from management.	The food service manager was unwilling to communicate ideas for improvement or become involved in improvements.
Meet with staff to go over the implementation plan.	Earl, Bonnie, C-1s, and C-3s	Tuesday (week 2)	Train staff on the new schedule.	Expected outcome met.
Implement changes for long-term care.	Earl and staff	Wednesday (week 2)	Implement changes for long-term care.	Changes implemented (1st iteration).
Data collection and analysis.	Earl	Wednesday (week 2)	Data collection and analysis.	Data collected and analyzed.
Make necessary changes to the plan.	Earl and staff	Thursday (week 2)	Fine-tune schedule of operations.	New schedule (second iteration) developed by stakeholders which adds Colin in hostess position.
Data collection and analysis.	Earl	Thursday (week 2)	Data collection and analysis.	Data collected and analyzed.
Present findings to staff and hospital management.	Earl	Friday (week 2)		

Effects Confirmation

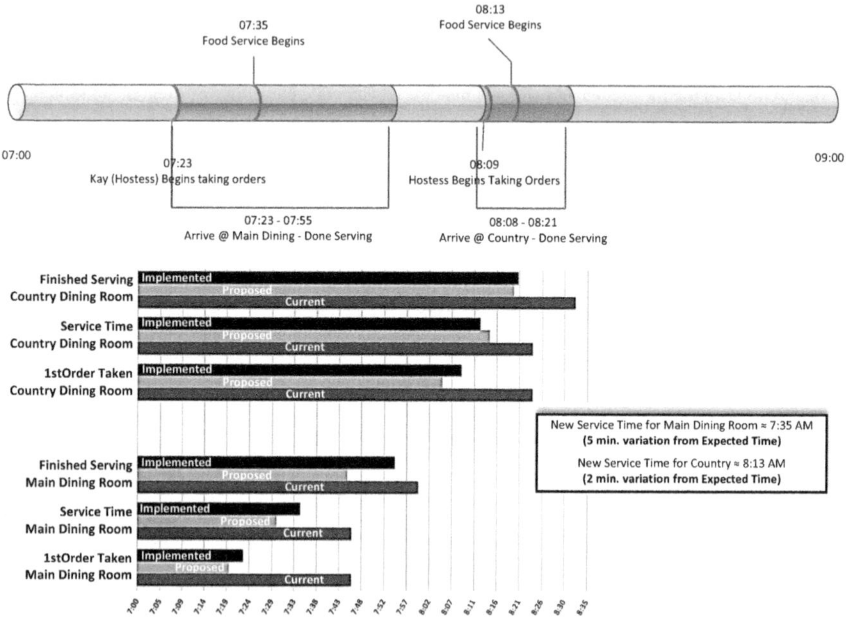

Results with Using "Hostess" to Take Orders & Serve Beverages:

General Observations:

- Using a "hostess" during meal service times allows for meals to be served at the designated times and provides more interaction with the residents.
- Data collected is for breakfast service; using a hostess for all meal services would have similar results.

SPECIAL NOTE: To make the neighborhood extended-care system work most effectively, the installation of dishwashers in the main kitchen is recommended. Some additional research indicates:

- A "Super-Capacity" or tall tub residential dishwasher holds up to 16 place settings (5-piece).
- Residential washers are available with sanitizing cycles to guarantee sanitary dishes.
- Both Country and Memory would be able to fit the dishes from each meal service into a residential model.
- A commercial, low-temp, under-counter dishwasher would be most beneficial for the main dining room.
- The dishes from Gaston and the main dining room could be done efficiently with a commercial under-counter machine. The low-temp washers do not require exhaust hoods.
- Some of the benefits associated with installing dishwashers in the pods are:
  - Dishes returning to Foodservice would be reduced to food pans (reducing the burden on the foodservice dish area).
  - The pods could be bused and cleaned by one staff member, eliminating the need of additional kitchen help to assist in bringing dishes to/from the foodservice kitchen.
  - Dishes and utensils for each pod could remain in place (no longer needing to be delivered and sorted).
  - The residents would benefit by having a more relaxed, less rushed dining experience.

Follow-Up Actions

Action	Responsible Party	To Complete Action	Due
Implement countermeasures for food service.	Bonnie, supervisors, and staff	Meet with all 1s and 3s to establish a standardized system to make long-term care easier to service.	ASAP
Schedule CNA breaks to coordinate with meal service.	Danielle, Colin, and Bonnie	Re-instate meal tickets to CNAs; determine if more/different types of food are needed to feed CNAs in pods at meal times.	ASAP
Have a lunch cart for Country brought down at 12:30; allow it to stay hot and not need to be loaded before serving Memory.	Bonnie and supervisors	Determine who is available to load hot food into the cart and deliver to Country.	In 4 weeks.
Install dishwashers in extended care.	Danielle, Colin, and Bonnie	Determine if plumbing is in place, acquire bids for one under-counter, low-temp machine and two residential models.	Once approval obtained.
Implement a new schedule utilizing dishwashers in pods.	Bonnie and staff	Reschedule 1s and 3s giving them time for dishes in pods instead up the kitchen dish-room; change busing schedule.	Upon dishwasher installation.

Case Study 9.5: Decrease Patient Waiting Times at a Clinic	Date: XX/XX/20XX
Sponsor: Clinic	Author: Karl

Background

- The DX Healthcare Clinic is open from 9 a.m. to noon and 1–5 p.m. Monday through Friday.
- There are currently three full-time providers that work 4 days a week in the clinic, and one of those days, they are also on call at the hospital ER.
- Locum tenens are used on Thursday and Friday for the on-call duty.
- A provider that has the on-call duty is responsible for any Emergency Room (ER) visits first and any walk-ins second.
- Each provider has a specific nurse who works with them unless the provider or nurse is gone.
- Each provider has two rooms that they use regularly to see their patients.
- There are two full-time receptionists who take all phone calls, check in patients, and gather charts.

Current Condition

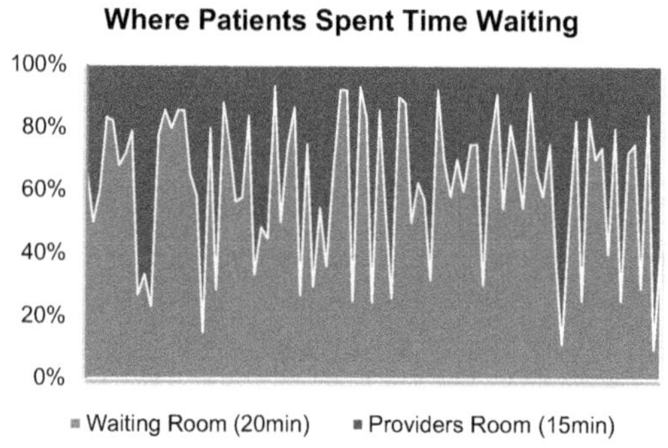

Current State Value Stream Map

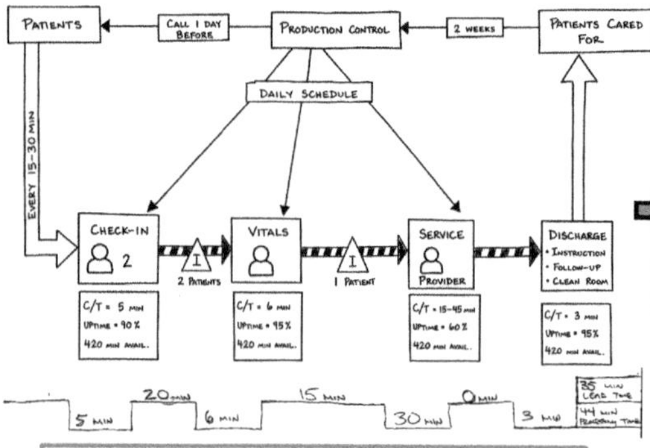

Average times spent waiting are represented in value-stream map as time spent in "inventory" by patient which is a non-value added activity

Current state summary:
Total processing time = 44 min per patient
Bottleneck rate = 2 patients/hour per provider
Critical work-in-process (W_0) = 1.46 patients

Average WIP = 2.6 patients
Current throughput is between the practical worst case and the worst case

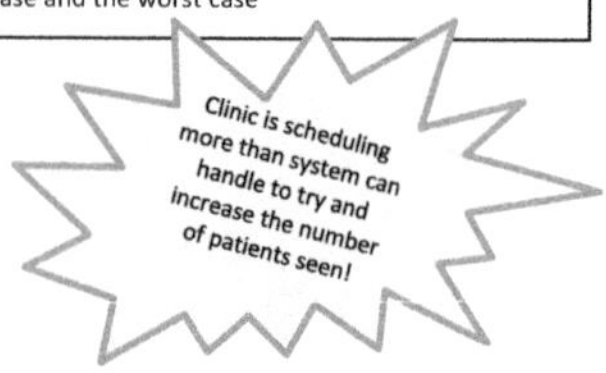

SMART Goal:	• *Decrease patient wait times in the clinic by at least 50%.*

Root Cause Analysis

Problem: Patients are waiting before seeing providers.

- **Why?** The provider not with the patient at the scheduled time.
  - **Why?** The provider not ready to see the patient.
    - **Why?** The provider busy with another patient.
      - **Why?** Not enough time scheduled for the patient.
        - **Why?** Trying to increase the number of patients seen.
  - **Why?** The patient not ready to be seen by the provider.
    - **Why?** Busy doing paperwork or getting vitals taken.
      - **Why?** Not taking into account the scheduling of patients.
        - **Why?** Scheduling for the time of the procedure rather than time for the entire patient process.

Proposed Countermeasures

- Make 30 minutes the minimum time for an appointment.
- Nurses and providers meet in morning to go over patient appointments for that day.

Target Condition:

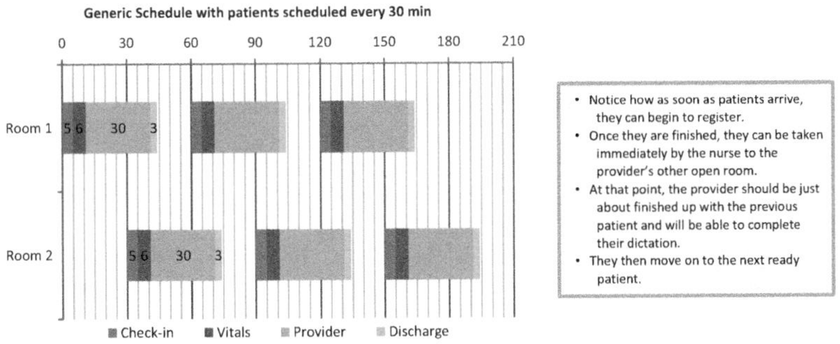

Implementation Plan

Action	Responsible	Deadline	Expected Outcome	Actual Outcome
Notify providers of findings and seek approval.	Karl	Tuesday (wk2)		
Notify the CEO and seek approval.	Karl	Tuesday (wk2)		
Meeting with entire staff from clinic and obtain feedback.	Karl	Tuesday (wk2)		
Change the scheduling system to 30-minutes blocks.	Karl and front desk	Wednesday (wk2)	Change schedule to 30 minutes blocks.	Schedule could not be changed; too close to appointment day.
Schedule future patients for 1 or 2 blocks.	Front desk	Wednesday (wk2)		
Specify time for nurses to meet with providers in morning.	Karl, nurses, providers	Wednesday (wk2)	Front desk print that day's schedule for nurse to go over with the provider.	Schedule printed, meeting took place.

Effects Confirmation

Since schedules could not be changed, I did a what-if study in which I compared what happened with the first five patients against what would have happened if they would have been scheduled in half an hour blocks.

Results:

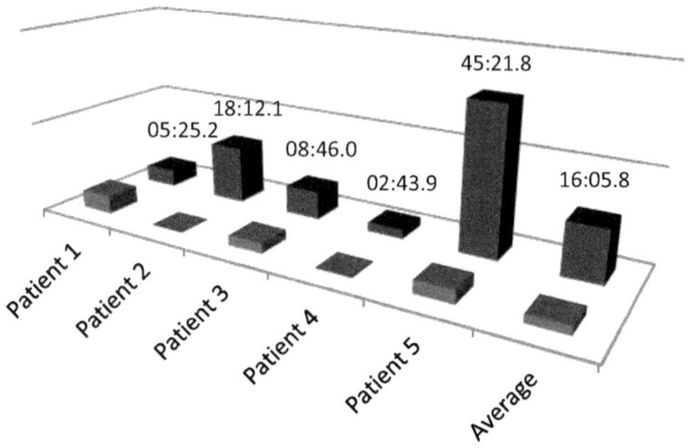

That would be an improvement of **85% for those patients' wait times**!

General recommendations and notes:

Any process will be limited by the bottleneck rate. Any improvements made at processes besides the bottleneck will be minimal compared to improvements made at the bottleneck.

That is why, for the clinic to start seeing significant improvement in the number of patients seen, improvements must be made at the provider process. This can be done by moving work tasks performed at the provider process to earlier in the patient process, increasing the number of providers, and/or increasing the availability of the providers throughout the day.

Once these improvements have been made, then patients can start to be scheduled more often, and improvements made in other parts of the clinic will have a more significant impact on the overall efficiency of the clinic.

Follow-Up Actions

Action	Responsible Party	To Complete Action	Due
Improve front desk operations.	Front desk, IT, Ian, and Lisa	Implement a new phone system with automated directing of calls and message machine.	In 2 weeks.
Improve the provider rate of seeing patients.	Providers, nurses, and Lisa	Have providers list what they need to complete the procedure and approximate time and begin to delegate some work tasks to nurses.	Ongoing next 4 weeks.
Increase provider availability.	Providers, Ian, and Lisa	Have providers work more often by opening clinic early, staying open longer, or working more days.	In 3 months.

Subject Index

H

I

J

K

L

M

N

O

P

Q

R

S

T

V

W

Case Study Index

K

L

M

N

O

P

Q

R

S

T

V

W

For Product Safety Concerns and Information please contact our EU representative GPSR@taylorandfrancis.com Taylor & Francis Verlag GmbH, Kaufingerstraße 24, 80331 München, Germany

Batch number: 10809472

Printed by Printforce, the Netherlands